Finite Element Analysis

M. Moatamedi
and
H. Khawaja

The International Society of Multiphysics
www.multiphysics.org

CRC Press
Taylor & Francis Group
Boca Raton London New York

CRC Press is an imprint of the
Taylor & Francis Group, an **informa** business
A SCIENCE PUBLISHERS BOOK

CRC Press
Taylor & Francis Group
6000 Broken Sound Parkway NW, Suite 300
Boca Raton, FL 33487-2742

First issued in paperback 2021

© 2018 by Taylor & Francis Group, LLC
CRC Press is an imprint of Taylor & Francis Group, an Informa business

No claim to original U.S. Government works

Version Date: 20180523

ISBN-13: 978-0-367-78102-6 (pbk)
ISBN-13: 978-1-138-32073-4 (hbk)

Visit the Taylor & Francis Web site at
http://www.taylorandfrancis.com

and the CRC Press Web site at
http://www.crcpress.com

Preface

Finite element analysis has become the most popular technique for studying engineering structures in detail. It is particularly useful whenever the complexity of the geometry or of the loading is such that alternative methods are inappropriate.

This book consists of seven chapters introducing the finite element method as a tool for the solution of practical engineering problems.

Chapter one introduces the aims and objectives of the book, history and perspective relating to the engineering problem, the fundamentals, numerical methods and application of the finite element method. This chapter also discusses the terminology associated with the finite element mesh.

Chapter two covers the matrix stiffness methods also referred to as displacement-based finite element method. It includes the formulation of simple bar element in local and global coordinate systems, assembly of bar elements and global stiffness matrix, loads and boundary conditions, a solution strategy supported by numerical examples. It also discusses error analysis and ill-conditioning, rigid body modes and mechanisms, symmetrical, antisymmetric and asymmetrical as well as thermal loadings.

Chapter three focuses on the finite element formulation of one-dimensional problems. It consists of the fundamental equations, shape function, algebraic and matrix form of finite element equations. It presents an example of element stiffness matrix for a 2-node bar with linear shape functions.

Chapter four provides the finite element formulation of two-dimensional problems covering the fundamental equations and finite element formulation for continuum for triangular and quadrilateral elements. It illustrates numerical examples and describes the restrictions on element formulations for completeness and compatibility.

Chapter five explains the computational implementation of finite element method introducing frontal and banded solution methodologies leading to the storage of the stiffness matrix and numerical integration by Gaussian quadrature.

Chapter six sets out the most commonly used elements namely beams, plates, shells and solids. It includes the impact of nodal degrees of freedom and boundary conditions in numerical examples.

Chapter seven details parametric element formulations. The focus is on isoperimetric bar element, isoperimetric four-node quadrilateral element, and isoperimetric eight-node quadrilateral element.

The book covers the principles of finite element analysis, including the mathematical fundamentals as required, to construct an appropriate finite element model of a physical system, and interpret the results of the analysis.

Contents

Chapter 1
Introduction

Finite element analysis has become the most popular technique for studying engineering structures in detail. It is particularly useful whenever the complexity of the geometry or of the loading is such that alternative methods are inappropriate. This book provides an introduction to the finite element method.

1.1 Book Aims and Objectives

- To introduce finite element analysis as a tool for the solution of practical engineering problems.
- To teach the principles of finite element analysis, including the mathematical fundamentals as required.
- To demonstrate how to construct an *appropriate* finite element model of a physical system, and how to interpret the results of the analysis.

1.2 History and Perspective

1.2.1 The engineering problem

There are three steps in the analytical solution of a physical problem:

i) Identify the variables.
ii) Formulate governing equations describing the physical system, including any constraints and boundary conditions.
iii) Solve the equations.

A most prudent fourth step is to validate the solutions with the experimental data.

Having identified the important variables, the governing equations for the system are formulated. Depending on the physical system under investigation, these might be based on principles of conservation of mass, conservation of momentum, conservation of energy, minimum potential energy, etc. At this stage it is almost always necessary to make simplifying assumptions in order to reduce the general equations to a form for which solutions can be sought.

Historically, many eminent mathematicians and scientists have formulated and solved governing equations for specific physical systems. It has often been necessary to invoke knowledge of empirical relationships between variables in order to generate acceptable solutions. Every engineering textbook is full of equations bearing the names of individuals who first developed the associated theories, and it is possible to gain a strong insight into the performance of most physical systems by performing calculations using these equations. Closed form solutions are available for many practical engineering problems, and until the late 1950s all engineering design analysis was based on such solutions. The introduction of the digital computer to engineering applications started a revolution in design analysis that is still gathering momentum. The ability of the computer to solve large systems of simultaneous equations led to the development of matrix methods for structural analysis problems, and these in turn led to the development of finite element analysis methods.

1.2.2 The finite element method

The finite element method is based on the premise that a complex structure can be broken down into a finite number of smaller pieces (elements), the behaviour of each of which is known or can be postulated. These elements might then be assembled, in some sense, to model the behaviour of the structure. Intuitively, this premise seems reasonable, but there are many important questions that need to be answered. In order to answer them, it is necessary to apply a degree of mathematical rigour to the development of finite element techniques. The approach that will be taken in this book is to develop the fundamental ideas and methodologies based on an intuitive engineering approach, and then to support them with appropriate mathematical proofs where necessary. It will rapidly become clear that the finite element method is an extremely powerful tool for the analysis of structures (and for other field problems), but that the volume of calculations required to solve all but the most trivial of them is such that the assistance of a computer is necessary.

It has been mentioned above that many questions arise concerning finite element analysis. Some of these questions are associated with the fundamental mathematical formulations, some with numerical solution techniques, and others with the practical application of the method. A short list of questions is presented below to give an indication of the issues that are important to the developers and users of this powerful analysis technique.

Fundamentals:

- Are there any restrictions on the postulated behaviour of the individual elements?
- Is convergence guaranteed? (i.e., does the accuracy of the solution improve as the structure is divided into more and more elements?)
- Can the intrinsic error in a finite element solution be estimated, or at least can it be bounded? Under what circumstances might the error be zero?

Numerical methods:

- The finite element method will require the solution of a potentially very large set of simultaneous equations. What algorithms are most appropriate for their solution, and how can the resources required be minimised?
- What restrictions on accuracy are imposed by numerical solutions—can they be estimated or bounded?

Application:

- How accurately does the model geometry need to reflect the actual geometry? Is it necessary to model local geometrical features such as fillet radii?
- How many elements, and of what type, are needed to analyse a particular structure under a particular set of loads?
- How might the resources required for an analysis be estimated?

In order to answer these questions, the engineer/analyst needs to understand both the nature and limitations of the finite element approximation and the fundamental behaviour of the structure. Misapplication of finite element analysis programs is most likely to arise when the analyst is ignorant of engineering phenomena.

1.3 The Finite Element Mesh: Terminology

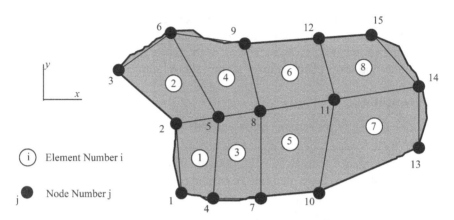

Figure 1.1 Description of Finite Element Mesh.

The physical domain (bounded by the bold line in Fig. 1.1) is broken down into discrete *elements*. At the boundaries of the domain, there is some approximation of the geometry unless the shape of the element edges corresponds exactly to the shape of the boundary.

The shapes of the elements are determined by the positions and connectivity of *nodes*. For simple triangular elements the minimum number of nodes required to define their shape and position is three (one at each corner). The simple quadrilaterals illustrated above are defined by the positions of the four corner nodes. More complex elements can be defined using more nodes: for example, an eight node quadrilateral can have curved edges defined by positions of mid-side nodes.

When the structure is loaded, each node can move from its original location. If the node is considered as a small solid particle there are six possible displacements in three dimensions. It can *translate* in any direction, and in general this translation will have a component along each of the three global axes. It can also *rotate*, and similarly the rotation has a component about each of the three global axes. Each of the components of translation and rotation is referred to as a *degree of freedom,* and in three dimensions each node, therefore, has six degrees of freedom. The number of *active* degrees of freedom at each node depends on the formulation of the elements connecting the nodes. If the area illustrated above represents a membrane in the xy plane, each node will have two active degrees of freedom: they are translation in the x direction (u), and translation in the y direction (v). The finite element model illustrated would, therefore, have $15 \times 2 = 30$ degrees of freedom.

Chapter 2

Matrix Stiffness Methods

The displacement-based finite element method is closely related to the matrix stiffness methods that were developed in the late 1950's in order to exploit the newly-arrived digital computers. Many of the procedures that are important in finite element analysis can be illustrated clearly and simply using these methods. In this chapter, the simple bar element is, developed and it is shown how a matrix equation can be constructed to represent the behaviour of an assembly of bars. Aspects of the numerical solution of this equation are also discussed.

2.1 The Simple Bar Element

2.1.1 Stiffness in co-ordinate system parallel to element axes

Figure 2.1 One-dimensional element.

Consider a bar with end points (nodes) i and j. Equations are required that relate the forces at the ends (X_i, X_j) to the displacements at the ends (u_i, u_j). Throughout this chapter, the bar on the variables indicates that they are specified in terms of the local co-ordinate system of the element.

By definition for a one-dimensional body, the strain (ε_x) is given by:

$$\varepsilon_x = \frac{d\bar{u}}{dx} \ .$$

By Hooke's Law in one dimension:

$$\varepsilon_x = \frac{\sigma_x}{E} \ .$$

For the bar illustrated above, with no body forces acting, the equilibrium requirement dictates that the force, F, at every cross-section must be constant. If positive F indicates a tensile force on the cross-section:

$$F = -\bar{X}_i = \bar{X}_j \ .$$

Eliminating strain from these equations, and substituting the basic definition for stress:

$$\frac{d\bar{u}}{dx} = \frac{\sigma_x}{E} = \frac{F}{A} \bullet \frac{1}{E} = \frac{-\bar{X}_i}{AE} \ .$$

Separating the variables:

$$\int_{\bar{u}_i}^{\bar{u}_j} d\bar{u} = \int_{x_i}^{x_j} \frac{-\bar{X}_i}{AE} \, dx \ .$$

Assuming that the cross-sectional area of the bar is constant along its length:

$$\bar{u}_j - \bar{u}_i = \frac{-\bar{X}_i}{AE}(x_j - x_i) \ .$$

If the length of the bar is L:

$$\bar{X}_i = \frac{AE}{L}(\bar{u}_i - \bar{u}_j) \ , \qquad \text{and similarly:} \qquad \bar{X}_j = \frac{AE}{L}(-\bar{u}_i + \bar{u}_j) \ .$$

Writing the last two equations in matrix form,

$$\begin{bmatrix} \bar{X}_i \\ \bar{X}_j \end{bmatrix} = \frac{AE}{L} \begin{bmatrix} 1 & -1 \\ -1 & 1 \end{bmatrix} \begin{bmatrix} \bar{u}_i \\ \bar{u}_j \end{bmatrix} \tag{2.1}$$

or:

$$\{\bar{X}\} = \left[\bar{k}\right]\{\bar{u}\} \tag{2.2}$$

where {X} and {u} are single column matrices, often referred to as force and displacement vectors, respectively, and [k] is a square matrix, in this case 2×2, representing the stiffness of the bar.

Some attributes of the stiffness matrix are of note:

   i) It is symmetrical about the main diagonal.
   ii) All of the entries on the main diagonal are positive.
   iii) The sums of the entries on each row and on each column is zero.
   iv) The determinant is zero (this follows from (iii)).

The relevance, importance and generality of these attributes will be discussed at a later stage.

The matrix equation developed above can be used to analyse uniaxial systems of bars under loads applied along their axes (for example a bar hanging vertically under its own weight, or a stepped bar subjected to an axial load). These applications are limited and such systems can readily be analysed using direct integration without any need for the finite element method. To have a practical value, it will be necessary to transform the stiffness matrix into global co-ordinates so that assemblies of bars forming pin-jointed frameworks might then be analysed.

2.1.2 Transformation to global co-ordinates

The stiffness matrix for a bar element orientated at an angle to the global axes in two dimensions is developed in this section. The extension to three dimensions simply creates more algebra but does not introduce any fundamental difficulties. Consider the bar element described in the previous section, this time orientated at an angle θ to the x axis.

8

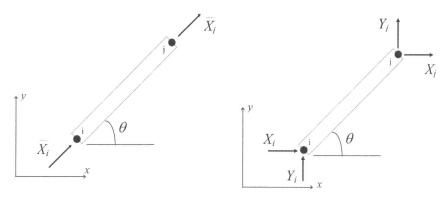

Figure 2.2 Global coordinate system.

The force applied at node i has components (X_i, Y_i) in global co-ordinates, and that applied at node j has components (X_j, Y_j). The force vector in global co-ordinates can be calculated from that in the local co-ordinates via multiplication by a rotation matrix as follows:

$$
\begin{bmatrix} X_i \\ Y_i \\ X_j \\ Y_j \end{bmatrix} = \begin{bmatrix} \cos\theta & -\sin\theta & 0 & 0 \\ \sin\theta & \cos\theta & 0 & 0 \\ 0 & 0 & \cos\theta & -\sin\theta \\ 0 & 0 & \sin\theta & \cos\theta \end{bmatrix} \begin{bmatrix} \overline{X}_i \\ 0 \\ \overline{X}_j \\ 0 \end{bmatrix}
$$

Writing the rotation matrix as [T], this equation can be expressed as:

$$\{X\} = [T]\{\overline{X}\} \tag{2.3}$$

Similarly, the displacements in the local axis system can be expressed in terms of their components in the global axes.

$$
\begin{bmatrix} \overline{u}_i \\ \overline{v}_i \\ \overline{u}_j \\ \overline{v}_j \end{bmatrix} = \begin{bmatrix} \cos\theta & \sin\theta & 0 & 0 \\ -\sin\theta & \cos\theta & 0 & 0 \\ 0 & 0 & \cos\theta & \sin\theta \\ 0 & 0 & -\sin\theta & \cos\theta \end{bmatrix} \begin{bmatrix} u_i \\ v_i \\ u_j \\ v_j \end{bmatrix}
$$

or, noting that the rotation matrix in this equation is the transpose of that in Equation (2.3):

$$\{\bar{u}\} = [T]^{T}\{u\} \tag{2.4}$$

Substituting Equations (2.2) and (2.4) into Equation (2.3):

$$\{X\} = [T]\{\bar{X}\} = [T][\bar{k}]\{\bar{u}\} = [T][\bar{k}][T]^{T}\{u\}$$

or:

$$\{X\} = [k]\{u\} \tag{2.5}$$

where $[k]$ is the stiffness matrix of the rod in global (two-dimensional) coordinates.

In expanded form:

$$
\begin{bmatrix} X_i \\ Y_i \\ X_j \\ Y_j \end{bmatrix} = \frac{AE}{L}
\begin{bmatrix}
\cos\theta & -\sin\theta & 0 & 0 \\
\sin\theta & \cos\theta & 0 & 0 \\
0 & 0 & \cos\theta & -\sin\theta \\
0 & 0 & \sin\theta & \cos\theta
\end{bmatrix}
\begin{bmatrix}
1 & 0 & -1 & 0 \\
0 & 0 & 0 & 0 \\
-1 & 0 & 1 & 0 \\
0 & 0 & 0 & 0
\end{bmatrix}
\begin{bmatrix}
\cos\theta & \sin\theta & 0 & 0 \\
-\sin\theta & \cos\theta & 0 & 0 \\
0 & 0 & \cos\theta & \sin\theta \\
0 & 0 & -\sin\theta & \cos\theta
\end{bmatrix}
\begin{bmatrix} u_i \\ v_i \\ u_j \\ v_j \end{bmatrix}
$$

Performing the matrix multiplications,

$$[k] = \frac{AE}{L}
\begin{bmatrix}
\cos^2\theta & \sin\theta\,\cos\theta & -\cos^2\theta & -\sin\theta\,\cos\theta \\
\sin\theta\,\cos\theta & \sin^2\theta & -\sin\theta\,\cos\theta & -\sin^2\theta \\
-\cos^2\theta & -\sin\theta\,\cos\theta & \cos^2\theta & \sin\theta\,\cos\theta \\
-\sin\theta\,\cos\theta & -\sin^2\theta & \sin\theta\,\cos\theta & \sin^2\theta
\end{bmatrix} \tag{2.6}$$

It is noted that all of the attributes of the stiffness matrix described earlier have been preserved under the transformation into global coordinates.

2.2 Assembly of Bar Elements—The Global Stiffness Matrix

The element stiffness matrix for a bar randomly orientated in two-dimensional space has been developed above. By combining the stiffness matrices of the individual members into a global stiffness matrix for the complete structure, any assembly of bars can now be analysed. Consider a pin-jointed framework consisting of five members as illustrated below.

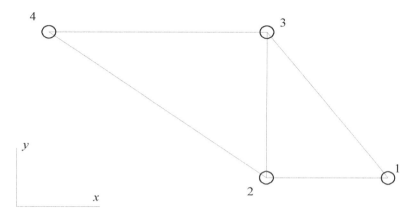

Figure 2.3 Pin-jointed framework.

The global stiffness matrix of the full assembly is required. This can be deduced from the individual element stiffness matrices. Consider the relationship of the global stiffness matrix with the force and displacement vectors.

$$
\begin{bmatrix} X_1 \\ Y_1 \\ X_2 \\ Y_2 \\ X_3 \\ Y_3 \\ X_4 \\ Y_4 \end{bmatrix} = \begin{bmatrix} k_{11} & k_{12} & k_{13} & k_{14} & k_{15} & k_{16} & k_{17} & k_{18} \\ k_{21} & k_{22} & k_{23} & k_{24} & k_{25} & k_{26} & k_{27} & k_{28} \\ k_{31} & k_{32} & k_{33} & k_{34} & k_{35} & k_{36} & k_{37} & k_{38} \\ k_{41} & k_{42} & k_{43} & k_{44} & k_{45} & k_{46} & k_{47} & k_{48} \\ k_{51} & k_{52} & k_{53} & k_{54} & k_{55} & k_{56} & k_{57} & k_{58} \\ k_{61} & k_{62} & k_{63} & k_{64} & k_{65} & k_{66} & k_{67} & k_{68} \\ k_{71} & k_{72} & k_{73} & k_{74} & k_{75} & k_{76} & k_{77} & k_{78} \\ k_{81} & k_{82} & k_{83} & k_{84} & k_{85} & k_{86} & k_{87} & k_{88} \end{bmatrix} \begin{bmatrix} u_1 \\ v_1 \\ u_2 \\ v_2 \\ u_3 \\ v_3 \\ u_4 \\ v_4 \end{bmatrix}
\tag{2.7}
$$

Now suppose that the framework is loaded in such a way that each of the displacements except for u_1 is zero, and that u_1 is unity (Figure 2.4). Then it is apparent that each of the entries in the first column of the global stiffness matrix is equal to the magnitude of the component of force in the force vector that is required to enforce this displacement pattern. These magnitudes are available from the individual element stiffness matrices.

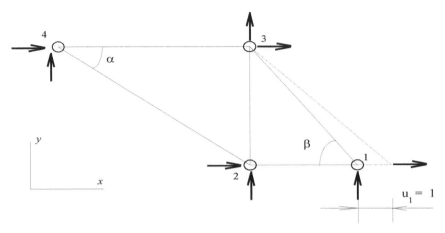

Figure 2.4 Loaded pin-jointed framework.

It is clear that none of the bars except for those directly connected to node 1 can be carrying any load, because for each of them the displacement at each end is zero and, therefore, the strain and, thus, the stress must also be zero. The bars that are connected to node 1 each carry a load that can be determined from their element stiffness matrices. Consider for example bar 1–3, inclined at an angle $(180-\beta)°$ to the global x axis. From Equations (2.5) and (2.6):

$$
\begin{bmatrix} X_1 \\ Y_1 \\ X_3 \\ Y_3 \end{bmatrix}_{1-3}
= \frac{AE}{L}
\begin{bmatrix}
\cos^2\beta & -\sin\beta\,\cos\beta & -\cos^2\beta & \sin\beta\,\cos\beta \\
-\sin\beta\,\cos\beta & \sin^2\beta & \sin\beta\cos\beta & -\sin^2\beta \\
-\cos^2\beta & \sin\beta\,\cos\beta & \cos^2\beta & -\sin\beta\,\cos\beta \\
\sin\beta\,\cos\beta & -\sin^2\beta & -\sin\beta\,\cos\beta & \sin^2\beta
\end{bmatrix}
\begin{bmatrix} 1 \\ 0 \\ 0 \\ 0 \end{bmatrix}
$$

$$(2.8)$$

Only the terms in the first column contribute to the forces on the element. Each of the forces on the element can be written down from this equation, and the magnitude of each is simply the magnitude of the equivalent entry in the first column of the stiffness matrix (not forgetting the AE/L term). Now, the total force applied in each direction at each node must be equal to the sum of those components of the forces that are in each bar. Therefore, each of the terms on the first column in the global stiffness matrix is the sum of the equivalent terms in the element stiffness matrix. The term on the main diagonal (k_{11}) will have contributions from every element that is connected to node 1, as will the term

k_{21} which also relates to the displacement of node 1. The remaining terms (k_{m1}) will have contributions only from those elements that connect both freedoms m and 1 (in a framework there will usually be only one such element).

Following a similar procedure, setting one degree of freedom equal to unity and all the remainder to zero, each column of the global stiffness matrix can be built-up. The procedure for assembly of the element stiffness matrices into the global stiffness matrix is thus established.

Note that the degree of redundancy of the framework has no impact on the matrix stiffness solution method. If an additional member connecting nodes 1 and 4 is introduced into the framework illustrated in Figure 2.4 the system is no longer statically determinate. Using this solution method, the member simply contributes additional terms to the rows and columns of the stiffness matrix that concern the degrees of freedom associated with nodes 1 and 4.

2.3 Loads and Boundary Conditions

Following the procedure outlined in the preceding section, a stiffness matrix for the complete structure can be assembled, and attention is now turned to the solution of Equation (2.7). The *ith* node in the assembly has two degrees of freedom (u_i and v_i), and each degree of freedom has an associated component of force (X_i or Y_i). In general, one of each associated pair (either the degree of freedom or its associated force component) has a specified numerical value, and the other is unknown. The basic approach in the matrix stiffness method (sometimes called the displacement method) is to treat the displacements as the primary unknowns and to solve for them first. Once the displacement solution is known, the unknown reaction forces can readily be computed from Equation (2.7).

In the most common types of problem encountered, the boundary conditions are a set of points or regions of the structure where some or all of the possible freedoms are constrained to be zero. It is at these points that the reaction forces are calculated. The forces at unconstrained nodes are zero unless the node is loaded.

2.4 A Solution Strategy

A strategy for the solution of the problem of an assembly of bars subjected to loads at their intersections can be defined based on the equations developed above.

i) Compute the individual stiffness matrices of each of the elements from Equation (2.6).

ii) Assemble the element stiffness matrices into the global stiffness matrix.

iii) Eliminate the rows and columns that correspond to zero-displacement boundary conditions.

iv) Solve the equations to find the displacements at each node.

v) Substitute the displacements into the original system equations to recover the reaction forces.

vi) Substitute the displacements into the equation relating strain within the element to the displacements at the nodes to calculate strains at points of interest.

vii) Calculate the stresses from the strains using Hooke's Law.

Step iv) involves the solution of a linear matrix equation, equivalent to a series of linear simultaneous equations. There are many ways in which this solution can be achieved. Traditionally, most finite element systems have used a procedure based on Gaussian elimination in some form, and the simplest form of this procedure is illustrated in the examples that follow. It consists of reducing the stiffness matrix to upper triangular form, performing the appropriate operations on the load vector at the same time, and then back-substituting to calculate each displacement in turn. A popular variation on this theme is based on the Choleski decomposition of the global stiffness matrix. Some modern systems use alternative, perhaps iterative, methods for the solution of the equations.

Step vi) involves a decision as to where in the element the strains will be evaluated. For bar elements, it seems obvious to evaluate strains, and hence stresses, at the nodes. Later in the book, elements will be introduced for which the evaluation of stress is not a trivial consideration.

2.5 Numerical Examples

2.5.1 Uniaxial system 1

The stepped bar illustrated below is constructed from steel (Young's modulus $E = 200000$ N/mm^2) and loaded as shown. The length of each segment of the bar is 500 mm. Calculate the load and the stress in each portion of the bar and the deflections at each step of the bar.

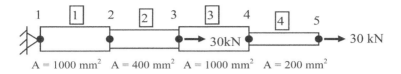

$$A = 1000 \text{ mm}^2 \quad A = 400 \text{ mm}^2 \quad A = 1000 \text{ mm}^2 \quad A = 200 \text{ mm}^2$$

For element 1:

$$[k]_{e1} = \frac{A_1 E_1}{L_1} \begin{bmatrix} 1 & -1 \\ -1 & 1 \end{bmatrix} = \frac{1000 \times 200000}{500} \begin{bmatrix} 1 & -1 \\ -1 & 1 \end{bmatrix}$$

$$\therefore \begin{Bmatrix} X_1 \\ X_2 \end{Bmatrix} = \begin{bmatrix} 400000 & -400000 \\ -400000 & 400000 \end{bmatrix} \begin{Bmatrix} u_1 \\ u_2 \end{Bmatrix}$$

$[k]$ is then formed for all the other elements.

For element 2:

$$[k]_{e2} = \frac{A_2 E_2}{L_2} \begin{bmatrix} 1 & -1 \\ -1 & 1 \end{bmatrix} = \frac{400 \times 200000}{500} \begin{bmatrix} 1 & -1 \\ -1 & 1 \end{bmatrix}$$

$$\begin{Bmatrix} X_2 \\ X_3 \end{Bmatrix} = \begin{bmatrix} 160000 & -160000 \\ -160000 & 160000 \end{bmatrix} \begin{Bmatrix} u_2 \\ u_3 \end{Bmatrix}$$

For element 3:

$$[k]_{e3} = \frac{A_3 E_3}{L_3} \begin{bmatrix} 1 & -1 \\ -1 & 1 \end{bmatrix} = \frac{1000 \times 200000}{500} \begin{bmatrix} 1 & -1 \\ -1 & 1 \end{bmatrix}$$

$$\begin{Bmatrix} X_3 \\ X_4 \end{Bmatrix} = \begin{bmatrix} 400000 & -400000 \\ -400000 & 400000 \end{bmatrix} \begin{Bmatrix} u_3 \\ u_4 \end{Bmatrix}$$

For element 4:

$$[k]_{e4} = \frac{A_4 E_4}{L_4} \begin{bmatrix} 1 & -1 \\ -1 & 1 \end{bmatrix} = \frac{200 \times 200000}{500} \begin{bmatrix} 1 & -1 \\ -1 & 1 \end{bmatrix}$$

$$\begin{Bmatrix} X_4 \\ X_5 \end{Bmatrix} = \begin{bmatrix} 80000 & -80000 \\ -80000 & 80000 \end{bmatrix} \begin{Bmatrix} u_4 \\ u_5 \end{Bmatrix}$$

The structure has five degrees of freedom, so the structure stiffness matrix will be 5 by 5.

Assembling the equations:

$$
\begin{Bmatrix} X_1 \\ X_2 \\ X_3 \\ X_4 \\ X_5 \end{Bmatrix} = 10^6 \begin{bmatrix} 0.4 & -0.4 & 0 & 0 & 0 \\ -0.4 & 0.4+0.16 & -0.16 & 0 & 0 \\ 0 & -0.16 & 0.16+0.4 & -0.4 & 0 \\ 0 & 0 & -0.4 & 0.4+0.08 & -0.08 \\ 0 & 0 & 0 & -0.08 & 0.08 \end{bmatrix} \begin{Bmatrix} u_1 \\ u_2 \\ u_3 \\ u_4 \\ u_5 \end{Bmatrix}
$$

$$
= 10^6 \begin{bmatrix} 0.4 & -0.4 & 0 & 0 & 0 \\ -0.4 & 0.56 & -0.16 & 0 & 0 \\ 0 & -0.16 & 0.56 & -0.4 & 0 \\ 0 & 0 & -0.4 & 0.48 & -0.08 \\ 0 & 0 & 0 & -0.08 & 0.08 \end{bmatrix} \begin{Bmatrix} u_1 \\ u_2 \\ u_3 \\ u_4 \\ u_5 \end{Bmatrix}
$$

Now substitute in the known forces and displacements:

$$
\begin{Bmatrix} X_1 \\ 0 \\ 30000 \\ 0 \\ 30000 \end{Bmatrix} = 10^6 \begin{bmatrix} 0.4 & -0.4 & 0 & 0 & 0 \\ -0.4 & 0.56 & -0.16 & 0 & 0 \\ 0 & -0.16 & 0.56 & -0.4 & 0 \\ 0 & 0 & -0.4 & 0.48 & -0.08 \\ 0 & 0 & 0 & -0.08 & 0.08 \end{bmatrix} \begin{Bmatrix} 0 \\ u_2 \\ u_3 \\ u_4 \\ u_5 \end{Bmatrix}
$$

Eliminate the rows and columns associated with zero displacements:

$$
\begin{Bmatrix} 0 \\ 30000 \\ 0 \\ 30000 \end{Bmatrix} = 10^6 \begin{bmatrix} 0.56 & -0.16 & 0 & 0 \\ -0.16 & 0.56 & -0.4 & 0 \\ 0 & -0.4 & 0.48 & -0.08 \\ 0 & 0 & -0.08 & 0.08 \end{bmatrix} \begin{Bmatrix} u_2 \\ u_3 \\ u_4 \\ u_5 \end{Bmatrix}
$$

This system of four linear simultaneous equations can be solved using Gaussian elimination. The entries below the main diagonal in each column are eliminated one by one.

16

Operating on the first column, there is one non-zero entry below the main diagonal. To eliminate this entry, multiply the first row by $(-0.16/0.56) = -0.2858$ and subtract it from the second row. The same operation must be performed on the force vector on the left-hand side of the equation. Thus, we get:

$$\begin{Bmatrix} 0 \\ 30000 \\ 0 \\ 30000 \end{Bmatrix} = 10^6 \begin{bmatrix} 0.56 & -0.16 & 0 & 0 \\ 0 & 0.5143 & -0.4 & 0 \\ 0 & -0.4 & 0.48 & -0.08 \\ 0 & 0 & -0.08 & 0.08 \end{bmatrix} \begin{Bmatrix} u_2 \\ u_3 \\ u_4 \\ u_5 \end{Bmatrix}$$

The row that remains unchanged, in this case the first, is the pivotal row.

Now operate on the second column to eliminate the entries below the main diagonal, with the second row becoming pivotal. Again, there is only one value in the current column below the main diagonal. This time the multiplication factor is $(-0.4/0.5143) = -0.7778$.

$$\begin{Bmatrix} 0 \\ 30000 \\ 23333 \\ 30000 \end{Bmatrix} = 10^6 \begin{bmatrix} 0.56 & -0.16 & 0 & 0 \\ 0 & 0.5143 & -0.4 & 0 \\ 0 & 0 & 0.1689 & -0.08 \\ 0 & 0 & -0.08 & 0.08 \end{bmatrix} \begin{Bmatrix} u_2 \\ u_3 \\ u_4 \\ u_5 \end{Bmatrix}$$

Finally, operate on the third column. The multiplication factor is now $(-0.08/0.1689) = -0.4737$.

$$\begin{Bmatrix} 0 \\ 30000 \\ 23333 \\ 41052 \end{Bmatrix} = 10^6 \begin{bmatrix} 0.56 & -0.16 & 0 & 0 \\ 0 & 0.5143 & -0.4 & 0 \\ 0 & 0 & 0.1689 & -0.08 \\ 0 & 0 & 0 & 0.0421 \end{bmatrix} \begin{Bmatrix} u_2 \\ u_3 \\ u_4 \\ u_5 \end{Bmatrix}$$

The displacement of Node 5 is now available directly from the last equation:

$$41052 = 0.0421 \times 10^6 \times u_5 \qquad\qquad u_5 = 0.9750 \text{ mm}$$

The displacement of Node 4 can be calculated from the penultimate equation:

$$23333 = 10^6 \times (0.1689 u_4 - 0.08 u_5) \qquad\qquad u_4 = 0.6000 \text{ mm}$$

Similarly, from the second and first equations respectively:

$u_3 = 0.5250$ mm   and $u_2 = 0.1500$ mm.

The above process is called *back-substitution*.

The reaction at Node 1 can be recovered from the first system equation that we temporarily discarded. Thus:

$X_1 = 10^6 \times (0.4u_1 - 0.4u_2) = -60000$ N

As would be hoped, this balances the total applied force exactly.

The strains can then be calculated from the displacements using the compatibility equation. Since the element is linear, i.e., a linear variation of displacement is assumed along each element, this is the difference in displacement of the two ends divided by the length of the element. The stresses can then be calculated using Hooke's law.

For element 1,

$$\varepsilon_x = \frac{u_2 - u_1}{L_1} = \frac{0.15 - 0}{500} = 300 \times 10^{-6}$$

$\sigma_x = E\varepsilon_x = 200000 \times 300 \times 10^{-6} = 60$ N/mm²

For element 2,

$$\varepsilon_x = \frac{u_3 - u_2}{L_2} = \frac{0.525 - 0.15}{500} = 750 \times 10^{-6}$$

$\sigma_x = E\varepsilon_x = 200000 \times 750 \times 10^{-6} = 150$ N/mm²

For element 3,

$$\varepsilon_x = \frac{u_4 - u_3}{L_3} = \frac{0.6 - 0.525}{500} = 150 \times 10^{-6}$$

$\sigma_x = E\varepsilon_x = 200000 \times 150 \times 10^{-6} = 30$ N/mm²

For element 4,

$$\varepsilon_x = \frac{u_5 - u_4}{L_4} = \frac{0.975 - 0.6}{500} = 750 \times 10^{-6}$$

$\sigma_x = E\varepsilon_x = 200000 \times 750 \times 10^{-6} = 150 \text{ N/mm}^2$

2.5.2 Uniaxial system 2

A platform of mass 500 kg sits symmetrically on top of a vertical steel pillar of cross-sectional area 0.002 m².

It is additionally supported by a vertical aluminium rod of cross-sectional area 0.005 m² attached to a rigid roof beam in line with the pillar.

If the height from floor to roof beam is 20 m and the platform is midway between, calculate the stresses in the steel and the aluminium when the platform is occupied by ten factory inspectors, each of mass 100 kg.

Neglect the weight of the pillar and the rod and assume that 1 kg mass produces a force of 10 N under the action of gravity.

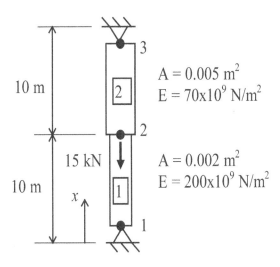

For element 1:

$$[k]_{e1} = \frac{A_1 E_1}{L_1}\begin{bmatrix} 1 & -1 \\ -1 & 1 \end{bmatrix} = \frac{0 \cdot 002 \times 200 \times 10^9}{10}\begin{bmatrix} 1 & -1 \\ -1 & 1 \end{bmatrix}$$

$$\therefore \begin{Bmatrix} X_1 \\ X_2 \end{Bmatrix} = 10^6\begin{bmatrix} 40 & -40 \\ -40 & 40 \end{bmatrix}\begin{Bmatrix} u_1 \\ u_2 \end{Bmatrix}$$

For element 2:

$$[k]_{e2} = \frac{A_2 E_2}{L_2}\begin{bmatrix} 1 & -1 \\ -1 & 1 \end{bmatrix} = \frac{0 \cdot 005 \times 70 \times 10^9}{10}\begin{bmatrix} 1 & -1 \\ -1 & 1 \end{bmatrix}$$

$$\begin{Bmatrix} X_2 \\ X_3 \end{Bmatrix} = 10^6\begin{bmatrix} 35 & -35 \\ -35 & 35 \end{bmatrix}\begin{Bmatrix} u_2 \\ u_3 \end{Bmatrix}$$

Assembling the element stiffness matrices to form the structure stiffness matrix:

$$\begin{Bmatrix} X_1 \\ X_2 \\ X_3 \end{Bmatrix} = 10^6\begin{bmatrix} 40 & -40 & 0 \\ -40 & (40+35) & -35 \\ 0 & -35 & 35 \end{bmatrix}\begin{Bmatrix} u_1 \\ u_2 \\ u_3 \end{Bmatrix}$$

Substitute the known forces and displacements:

$$\begin{Bmatrix} X_1 \\ -15000 \\ X_3 \end{Bmatrix} = 10^6\begin{bmatrix} 40 & -40 & 0 \\ -40 & 75 & -35 \\ 0 & -35 & 35 \end{bmatrix}\begin{Bmatrix} 0 \\ u_2 \\ 0 \end{Bmatrix}$$

Eliminate the rows and columns corresponding to zero displacements:

$$-15000 = 75 \times 10^6 \times u_2$$

$$u_2 = -0.0002 \text{ m} = -0.2 \text{ mm}$$

The reaction forces at Points 1 and 3 can be recovered from the full set of system equations:

$X_1 = 10^6 \,(40 \times 0 - 40 \times u_2 + 0 \times 0) = 8000 \text{ N}$

$X_3 = 10^6 \,(0 \times 0 - 35 \times u_2 + 35 \times 0) = 7000 \text{ N}$

The strains can be calculated from the displacements, and then the stresses can be calculated from the strains:

Element 1

$$\varepsilon_x = \frac{(u_2 - u_1)_1}{l_1} = \frac{-0.0002 - 0}{10} = -20 \times 10^{-6}$$

$$\sigma_x = E\varepsilon_x = 200 \times 10^9 \times \left(-20 \times 10^{-6}\right)$$

$$= -4 \times 10^6 \text{ N/m}^2$$

Element 2

$$\varepsilon_x = \frac{(u_3 - u_2)_2}{l_2} = \frac{0 + 0.0002}{10} = 20 \times 10^{-6}$$

$$\sigma_x = E\varepsilon_x = 70 \times 10^9 \times \left(20 \times 10^{-6}\right)$$

$$= 1.4 \times 10^6 \text{ N/m}^2$$

2.5.3 Pin-jointed framework 1

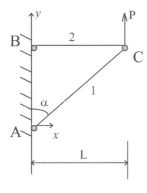

Both members: Area $A = 100 \text{ mm}^2$,
Young's modulus
$E = 70 \text{ kN/mm}^2$

$\alpha = 40°$

$L = 750 \text{ mm}$

$P = 10 \text{ kN}$

Find the displacement for the node C. Which can be obtained using the following analytical solutions

$$u_C = -\frac{PL}{AE}\tan\alpha$$

$$v_C = \frac{PL}{AE}\left(\tan^2\alpha + \frac{1}{\cos^2\alpha\sin\alpha}\right)$$

Element 1: $\theta_1 = 50°$

$A_1 = 100$ mm² $\qquad E_1 = 70 \times 10^3$ N/mm²

$$L_1 = \frac{750}{\sin 40°} = 1166.79 \text{ mm}$$

$$[k]_1 = \frac{100\times70\times10^3}{1166.79}\begin{bmatrix} 0.4132 & 0.4924 & -0.4132 & -0.4924 \\ 0.4924 & 0.5868 & -0.4924 & -0.5868 \\ -0.4132 & -0.4924 & 0.4132 & 0.4924 \\ -0.4924 & -0.5868 & 0.4924 & 0.5868 \end{bmatrix}$$

$$\therefore [k]_1 = 10^3\begin{bmatrix} 2.4789 & 2.9541 & -2.4789 & -2.9541 \\ 2.9541 & 3.5204 & -2.9541 & -3.5204 \\ -2.4789 & -2.9541 & 2.4789 & 2.9541 \\ -2.9541 & -3.5204 & 2.9541 & 3.5204 \end{bmatrix}$$

Element 2: $\theta_2 = 0°$

$A_2 = 100$ mm²

$E_2 = 70 \times 10^3$ N/mm²

$L_2 = 750$ mm

$$[k]_2 = \frac{100 \times 70 \times 10^3}{750} \begin{bmatrix} 1 & 0 & -1 & 0 \\ 0 & 0 & 0 & 0 \\ -1 & 0 & 1 & 0 \\ 0 & 0 & 0 & 0 \end{bmatrix}$$

$$\therefore [k]_2 = 10^3 \begin{bmatrix} 9.3333 & 0 & -9.3333 & 0 \\ 0 & 0 & 0 & 0 \\ -9.3333 & 0 & 9.3333 & 0 \\ 0 & 0 & 0 & 0 \end{bmatrix}$$

after assembling the element stiffness matrices, we get:

$$\begin{Bmatrix} X_1 \\ Y_1 \\ X_2 \\ Y_2 \\ X_3 \\ Y_3 \end{Bmatrix} = 10^3 \begin{bmatrix} 2.4789 & 2.9541 & 0 & 0 & -2.4789 & -2.9541 \\ 2.9541 & 3.5204 & 0 & 0 & -2.9541 & -3.5204 \\ 0 & 0 & 9.3333 & 0 & -9.3333 & 0 \\ 0 & 0 & 0 & 0 & 0 & 0 \\ -2.4789 & -2.9541 & -9.3333 & 0 & 11.8122 & 2.9541 \\ -2.9541 & -3.5204 & 0 & & 2.9541 & 3.5204 \end{bmatrix} \begin{Bmatrix} u_1 \\ v_1 \\ u_2 \\ v_2 \\ u_3 \\ v_3 \end{Bmatrix}$$

as the full set of equations, where 1, 2 and 3 refers to nodes A, B and C respectively. By substituting the nodal displacements and forces,

$$\begin{Bmatrix} X_1 \\ Y_1 \\ X_2 \\ Y_2 \\ 0 \\ 10^4 \end{Bmatrix} = 10^3 \begin{bmatrix} 2.4789 & 2.9541 & 0 & 0 & -2.4789 & -2.9541 \\ 2.9541 & 3.5204 & 0 & 0 & -2.9541 & -3.5204 \\ 0 & 0 & 9.3333 & 0 & -9.3333 & 0 \\ 0 & 0 & 0 & 0 & 0 & 0 \\ -2.4789 & -2.9541 & -9.3333 & 0 & 11.8122 & 2.9541 \\ -2.9541 & -3.5204 & 0 & & 2.9541 & 3.5204 \end{bmatrix} \begin{Bmatrix} 0 \\ 0 \\ 0 \\ 0 \\ u_3 \\ v_3 \end{Bmatrix}$$

which results following equations,

$$X_1 = -2478.9u_3 - 2954.1v_3$$

$$Y_1 = -2954.1u_3 - 3520.4v_3$$

$X_2 = -9333.3u_3$

$Y_2 = 0$

$0 = 11812.2u_3 + 2954.1v_3$

$10000 = 2954.1u_3 + 3520.4v_3$

By solving the last two equations simultaneously,

$u_3 = -0.8991$ *mm*

$v_3 = 3.595$ *mm*

These values can be compared with the analytical solution of u_C and v_C provided earlier.

2.5.4 Redundant pin-jointed framework

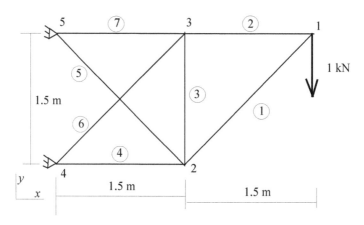

Figure 2.5 Redundant pin-jointed framework.

A space-frame internal wing structure of a light kit-form aircraft can be idealised as a pin-jointed framework as illustrated above. The framework is connected to a very stiff piece of structure at nodes 4 and 5, and the displacements of each of these nodes can be considered to be zero in both x and y directions. Each of the members in the framework has an extensional stiffness $AE/L = 10$ kN/mm.

Calculate the displacement of each of the joints and the force in each member when a load of 1 kN is applied in the y direction at node 1.

The first step in the solution procedure is to write down the element stiffness matrices. For each bar, orientated at an angle θ (measured anti-clockwise) to the global x axis, the force/displacement equation can be derived from Equations 2.5 and 2.6. For example, for bar 1:

$$
\begin{bmatrix} X_1 \\ Y_1 \\ X_2 \\ Y_2 \end{bmatrix}_1 = \left(\frac{AE}{L}\right)_{\text{element 1}}
\begin{bmatrix}
\cos^2\theta & \sin\theta\,\cos\theta & -\cos^2\theta & -\sin\theta\,\cos\theta \\
\sin\theta\,\cos\theta & \sin^2\theta & -\sin\theta\,\cos\theta & -\sin^2\theta \\
-\cos^2\theta & -\sin\theta\,\cos\theta & \cos^2\theta & \sin\theta\,\cos\theta \\
-\sin\theta\,\cos\theta & -\sin^2\theta & \sin\theta\,\cos\theta & \sin^2\theta
\end{bmatrix}
\begin{bmatrix} u_1 \\ v_1 \\ u_2 \\ v_2 \end{bmatrix}
$$

Stiffness matrices for each of the seven bar elements are written below.

Element 1 AE/L = 10000 N/mm: $\theta = -135°$: $\sin\theta = -0.7071$ $\cos\theta = -0.7071$

$$
\begin{bmatrix} X_1 \\ Y_1 \\ X_2 \\ Y_2 \end{bmatrix}_1 =
\begin{bmatrix}
5000 & 5000 & -5000 & -5000 \\
5000 & 5000 & -5000 & -5000 \\
-5000 & -5000 & 5000 & 5000 \\
-5000 & -5000 & 5000 & 5000
\end{bmatrix}
\begin{bmatrix} u_1 \\ v_1 \\ u_2 \\ v_2 \end{bmatrix}
$$

Element 2 AE/L = 10000 N/mm: $\theta = 180°$: $\sin\theta = 0$ $\cos\theta = -1$

$$
\begin{bmatrix} X_1 \\ Y_1 \\ X_3 \\ Y_3 \end{bmatrix}_2 =
\begin{bmatrix}
10000 & 0 & -10000 & 0 \\
0 & 0 & 0 & 0 \\
-10000 & 0 & 10000 & 0 \\
0 & 0 & 0 & 0
\end{bmatrix}
\begin{bmatrix} u_1 \\ v_1 \\ u_3 \\ v_3 \end{bmatrix}
$$

Element 3 AE/L = 10000 N/mm: $\theta = 90°$: $\sin\theta = 1$ $\cos\theta = -0.7071$

$$
\begin{bmatrix} X_2 \\ Y_2 \\ X_3 \\ Y_3 \end{bmatrix}_3 =
\begin{bmatrix}
0 & 0 & 0 & 0 \\
0 & 10000 & 0 & -10000 \\
0 & 0 & 0 & 0 \\
0 & -10000 & 0 & 10000
\end{bmatrix}
\begin{bmatrix} u_2 \\ v_2 \\ u_3 \\ v_3 \end{bmatrix}
$$

Element 4 AE/L = 10000 N/mm: $\theta = 180°$: $\sin\theta = 0$ $\cos\theta = -1$

$$
\begin{bmatrix} X_2 \\ Y_2 \\ X_4 \\ Y_4 \end{bmatrix}_4 =
\begin{bmatrix} 10000 & 0 & -10000 & 0 \\ 0 & 0 & 0 & 0 \\ -10000 & 0 & 10000 & 0 \\ 0 & 0 & 0 & 0 \end{bmatrix}
\begin{bmatrix} u_2 \\ v_2 \\ u_4 \\ v_4 \end{bmatrix}
$$

Element 5 AE/L = 10000 N/mm: $\theta = 135°$: $\sin\theta = 0.7071$ $\cos\theta = -0.7071$

$$
\begin{bmatrix} X_2 \\ Y_2 \\ X_5 \\ Y_5 \end{bmatrix}_5 =
\begin{bmatrix} 5000 & -5000 & -5000 & 5000 \\ -5000 & 5000 & 5000 & -5000 \\ -5000 & 5000 & 5000 & -5000 \\ 5000 & -5000 & -5000 & 5000 \end{bmatrix}
\begin{bmatrix} u_2 \\ v_2 \\ u_5 \\ v_5 \end{bmatrix}
$$

Element 6 AE/L = 10000 N/mm: $\theta = -135°$: $\sin\theta = -0.7071$ $\cos\theta = -0.7071$

$$
\begin{bmatrix} X_3 \\ Y_3 \\ X_4 \\ Y_4 \end{bmatrix}_6 =
\begin{bmatrix} 5000 & 5000 & -5000 & -5000 \\ 5000 & 5000 & -5000 & -5000 \\ -5000 & -5000 & 5000 & 5000 \\ -5000 & -5000 & 5000 & 5000 \end{bmatrix}
\begin{bmatrix} u_3 \\ v_3 \\ u_4 \\ v_4 \end{bmatrix}
$$

Element 7 AE/L = 10000 N/mm: $\theta = 180°$: $\sin\theta = 0$ $\cos\theta = -1$

$$
\begin{bmatrix} X_3 \\ Y_3 \\ X_5 \\ Y_5 \end{bmatrix}_7 =
\begin{bmatrix} 10000 & 0 & -10000 & 0 \\ 0 & 0 & 0 & 0 \\ -10000 & 0 & 10000 & 0 \\ 0 & 0 & 0 & 0 \end{bmatrix}
\begin{bmatrix} u_3 \\ v_3 \\ u_5 \\ v_5 \end{bmatrix}
$$

The element stiffness matrices are now assembled into a global stiffness matrix for the structure.

$$
\begin{bmatrix} X_1 \\ Y_1 \\ X_2 \\ Y_2 \\ X_3 \\ Y_3 \\ X_4 \\ Y_4 \\ X_5 \\ Y_5 \end{bmatrix}
=
\begin{bmatrix}
5000+10000 & 5000+0 & -5000 & -5000 & -10000 & 0 & 0 & 0 & 0 & 0 \\
 & 5000+0 & -5000 & -5000 & 0 & 0 & 0 & 0 & 0 & 0 \\
 & & 5000+0+10000+5000 & 5000+0+0-5000 & 0 & 0 & -10000 & 0 & -5000 & 5000 \\
 & & & 5000+10000+0+5000 & 0 & -10000 & 0 & 0 & 5000 & -5000 \\
 & & & & 10000+0+5000+10000 & 0+0+5000+0 & -5000 & -5000 & -10000 & 0 \\
 & & & & & 0+10000+5000+0 & -5000 & -5000 & 0 & 0 \\
 & & & & & & 0+10000+5000 & 0+5000 & 0 & 0 \\
 & & symmetrical & & & & & 0+5000 & 0 & 0 \\
 & & & & & & & & 5000+10000 & -5000+0 \\
 & & & & & & & & & 5000+0
\end{bmatrix}
\begin{bmatrix} u_1 \\ v_1 \\ u_2 \\ v_2 \\ u_3 \\ v_3 \\ u_4 \\ v_4 \\ u_5 \\ v_5 \end{bmatrix}
$$

Note that there are as many contributors to each of the entries on the main diagonal as there are members joining at the node under consideration. Adding the contributions together, the global stiffness matrix is calculated. At this stage, the known values of force and displacement at the nodes are entered into the force and displacement vectors.

$$
\begin{bmatrix} 0 \\ -1000 \\ 0 \\ 0 \\ 0 \\ 0 \\ X_4 \\ Y_4 \\ X_5 \\ Y_5 \end{bmatrix}
=
\begin{bmatrix}
15000 & 5000 & -5000 & -5000 & -10000 & 0 & 0 & 0 & 0 & 0 \\
5000 & 5000 & -5000 & -5000 & 0 & 0 & 0 & 0 & 0 & 0 \\
-5000 & -5000 & 20000 & 0 & 0 & 0 & -10000 & 0 & -5000 & 5000 \\
-5000 & -5000 & 0 & 20000 & 0 & -10000 & 0 & 0 & 5000 & -5000 \\
-10000 & 0 & 0 & 0 & 25000 & 5000 & -5000 & -5000 & -10000 & 0 \\
0 & 0 & 0 & -10000 & 5000 & 15000 & -5000 & -5000 & 0 & 0 \\
0 & 0 & -10000 & 0 & -5000 & -5000 & 15000 & 5000 & 0 & 0 \\
0 & 0 & 0 & 0 & -5000 & -5000 & 5000 & 5000 & 0 & 0 \\
0 & 0 & -5000 & 5000 & -10000 & 0 & 0 & 0 & 15000 & -5000 \\
0 & 0 & 5000 & -5000 & 0 & 0 & 0 & 0 & -5000 & 5000
\end{bmatrix}
\begin{bmatrix} u_1 \\ v_1 \\ u_2 \\ v_2 \\ u_3 \\ v_3 \\ 0 \\ 0 \\ 0 \\ 0 \end{bmatrix}
$$

27

The first six equations contain the six unknown components of displacement (degrees of freedom) and no unknown forces. The equations are all independent and so they can be solved for displacement.

$$
\begin{bmatrix} 0 \\ -1000 \\ 0 \\ 0 \\ 0 \\ 0 \end{bmatrix} = \begin{bmatrix} 15000 & 5000 & -5000 & -5000 & -10000 & 0 \\ 5000 & 5000 & -5000 & -5000 & 0 & 0 \\ -5000 & -5000 & 20000 & 0 & 0 & 0 \\ -5000 & -5000 & 0 & 20000 & 0 & -10000 \\ -10000 & 0 & 0 & 0 & 25000 & 5000 \\ 0 & 0 & 0 & -10000 & 5000 & 15000 \end{bmatrix} \begin{bmatrix} u_1 \\ v_1 \\ u_2 \\ v_2 \\ u_3 \\ v_3 \end{bmatrix}
$$

There are many approaches to the solution of these simultaneous equations. One approach is to use Gaussian elimination to reduce the matrix to upper triangular. To eliminate the entry in the jth column of the ith row ($i > j$), the standard technique is to multiply the jth row by a factor k_{ij}/k_{jj}, and then to subtract it from row i. For example, to eliminate the entry -10000 from the fifth row of the first column in the above matrix, the first row is multiplied by the factor ($-10000/15000$) and then subtracted from the fifth row.

Eliminating all of the entries below the leading diagonal in the first column in like manner gives:

Multiplication
Factor

$$
\begin{bmatrix} 0 \\ -1000 \\ 0 \\ 0 \\ 0 \\ 0 \end{bmatrix} = \begin{bmatrix} 15000 & 5000 & -5000 & -5000 & -10000 & 0 \\ 0 & 3333 & -3333 & -3333 & 3333 & 0 \\ 0 & -3333 & 18333 & -1667 & -3333 & 0 \\ 0 & -3333 & -1667 & 18333 & -3333 & -10000 \\ 0 & 3333 & -3333 & -3333 & 18333 & 5000 \\ 0 & 0 & 0 & -10000 & 5000 & 15000 \end{bmatrix} \begin{bmatrix} u_1 \\ v_1 \\ u_2 \\ v_2 \\ u_3 \\ v_3 \end{bmatrix}
\quad
\begin{matrix} \\ 1/3 \\ -1/3 \\ -1/3 \\ -2/3 \\ 0 \end{matrix}
$$

Eliminating entries below the leading diagonal in the second column:

Multiplication
Factor

$$
\begin{bmatrix} 0 \\ -1000 \\ -1000 \\ -1000 \\ 1000 \\ 0 \end{bmatrix} = \begin{bmatrix} 15000 & 5000 & -5000 & -5000 & -10000 & 0 \\ 0 & 3333 & -3333 & -3333 & 3333 & 0 \\ 0 & 0 & 15000 & -5000 & 0 & 0 \\ 0 & 0 & -5000 & 15000 & 0 & -10000 \\ 0 & 0 & 0 & 0 & 15000 & 5000 \\ 0 & 0 & 0 & -10000 & 5000 & 17071 \end{bmatrix} \begin{bmatrix} u_1 \\ v_1 \\ u_2 \\ v_2 \\ u_3 \\ v_3 \end{bmatrix}
\qquad
\begin{matrix} \\ \\ -1 \\ -1 \\ 1 \\ 0 \end{matrix}
$$

Eliminating entries below the leading diagonal in the third column:

Multiplication
Factor

$$
\begin{bmatrix} 0 \\ -1000 \\ -1000 \\ -1333 \\ 1000 \\ 0 \end{bmatrix} = \begin{bmatrix} 15000 & 5000 & -5000 & -5000 & -10000 & 0 \\ 0 & 3333 & -3333 & -3333 & 3333 & 0 \\ 0 & 0 & 15000 & -5000 & 0 & 0 \\ 0 & 0 & 0 & 13333 & 0 & -10000 \\ 0 & 0 & 0 & 0 & 15000 & 5000 \\ 0 & 0 & 0 & -10000 & 5000 & 17071 \end{bmatrix} \begin{bmatrix} u_1 \\ v_1 \\ u_2 \\ v_2 \\ u_3 \\ v_3 \end{bmatrix}
\qquad
\begin{matrix} \\ \\ \\ -1/3 \\ 0 \\ 0 \end{matrix}
$$

Eliminating entries below the leading diagonal in the fourth column:

Multiplication
Factor

$$
\begin{bmatrix} 0 \\ -1000 \\ -1000 \\ -1333 \\ 1000 \\ -1000 \end{bmatrix} = \begin{bmatrix} 15000 & 5000 & -5000 & -5000 & -10000 & 0 \\ 0 & 3333 & -3333 & -3333 & 3333 & 0 \\ 0 & 0 & 15000 & -5000 & 0 & 0 \\ 0 & 0 & 0 & 13333 & 0 & -10000 \\ 0 & 0 & 0 & 0 & 15000 & 5000 \\ 0 & 0 & 0 & 0 & 5000 & 7500 \end{bmatrix} \begin{bmatrix} u_1 \\ v_1 \\ u_2 \\ v_2 \\ u_3 \\ v_3 \end{bmatrix}
\qquad
\begin{matrix} \\ \\ \\ \\ 0 \\ -3/4 \end{matrix}
$$

Finally, eliminating entries below the leading diagonal in the fifth column:

Multiplication
Factor

$$
\begin{bmatrix} 0 \\ -1000 \\ -1000 \\ -1333 \\ 1000 \\ -1333 \end{bmatrix} = \begin{bmatrix} 15000 & 5000 & -5000 & -5000 & -10000 & 0 \\ 0 & 3333 & -3333 & -3333 & 3333 & 0 \\ 0 & 0 & 15000 & -5000 & 0 & 0 \\ 0 & 0 & 0 & 13333 & 0 & -10000 \\ 0 & 0 & 0 & 0 & 15000 & 5000 \\ 0 & 0 & 0 & 0 & 0 & 5833 \end{bmatrix} \begin{bmatrix} u_1 \\ v_1 \\ u_2 \\ v_2 \\ u_3 \\ v_3 \end{bmatrix}
$$

1/3

The solution for the displacements can now be determined by back-substitution.

$$v_3 = \frac{-1333}{5833} = -0.2286 \text{ mm}$$

$$u_3 = \frac{1}{15000}\left(1000 - 5000v_3\right) = 0.1429 \text{ mm}$$

$$v_2 = \frac{1}{13333}\left(-1333 - 0u_3 + 10000v_3\right) = -0.2714 \text{ mm}$$

$$u_2 = \frac{1}{15000}\left(-1000 + 5000v_2 - 0u_3 - 0v_3\right) = -0.1571 \text{ mm}$$

$$v_1 = \frac{1}{3333}\left(-1000 + 3333u_2 + 3333v_2 - 3333u_3 - 0v_3\right) = -0.8714 \text{ mm}$$

$$u_1 = \frac{1}{15000}\left(0 - 5000v_1 + 5000u_2 + 5000v_2 + 10000u_3 - 0v_3\right) = 0.2429 \text{ mm}$$

The reaction forces at the points of restraint can be recovered from the global equation.

$$X_4 = 0u_1 + 0v_1 - 10000u_2 + 0v_2 - 5000u_3 - 5000v_3 = 2000 \text{ N}$$

$$Y_4 = 0u_1 + 0v_1 + 0u_2 + 0v_2 - 5000u_3 - 5000v_3 = 428.5 \text{ N}$$

$X_5 = 0u_1 + 0v_1 - 5000u_2 + 5000v_2 - 10000u_3 + 0v_3 = -2000$ N

$Y_5 = 0u_1 + 0v_1 + 5000u_2 - 5000v_2 + 0u_3 + 0v_3 = 571.5$ N

2.6 Error Analysis and Ill-conditioning

2.6.1 Theory

In each of the numerical examples studied in the previous section, a solution for the nodal displacements has been found. The solutions will not in general be exact because the computations are carried out to a finite precision, and truncation and rounding errors will be present. There are many ways in which the accuracy of the solution might be measured. It is important to understand the principles of error estimation, and to be able to recognise systems in which numerical errors are likely to arise. Two residual vectors are defined as follows.

The residual displacement vector, $\{r\}$, is defined as the difference between the exact solution and the numerical solution of the governing equation.

$$\{r\} = \{u\} - \{u\}_{sol} \tag{2.9}$$

The exact solution is not known *a priori*, therefore, some means of estimating the residual displacement vector is required.

The residual force vector, $\{R\}$, is defined as the difference between the original applied force vector and the force vector calculated from the stiffness matrix multiplied by the numerical nodal displacement solution.

$$\{R\} = \{X\} - [k]\{u\}_{sol} \tag{2.10}$$

The residual force vector can readily be calculated, once the numerical solution for the displacement vector is known, using one matrix multiplication and one subtraction. The computation of the residual force vector should be carried out using a higher precision than that used for the analysis solution. For example, if the analysis is carried out using single precision arithmetic, the residual force vector should be calculated using double precision. The magnitude of the residual force vector can be compared with that of the applied force vector using any appropriate vector norm. It is clear that the $\{R\}$ must be small relative to $\{X\}$ if the solution is accurate. It will now be shown that this is a necessary but not a sufficient condition for accuracy of the displacement solution.

The relationship between the residual displacement vector and the residual force vector can be calculated as follows.

$$\{R\} = [k]\{u\} - [k](\{u\} - \{r\})$$

$$\{r\} = [k]^{-1}\{R\} \qquad (2.11)$$

It is apparent that the residual displacement vector might be large, even when $\{R\}$ is small, when the elements of $[k]^{-1}$ are large. It can be shown that errors in the numerical solution are likely when the *condition number*, cond(k), of the stiffness matrix is large. The condition number is defined as the ratio of the largest to the smallest eigenvalue of the stiffness matrix.

2.6.2 Numerical examples

2.6.2.1 Error analysis of Example 2.5.4

For example, 2.5.4, the displacement solution is:

$$\{u\}_{sol} = \begin{Bmatrix} 0\cdot 2429 \\ -0\cdot 8714 \\ -0\cdot 1571 \\ -0\cdot 2714 \\ 0\cdot 1429 \\ -0\cdot 2286 \end{Bmatrix}$$

The force vector based on this solution is:

$$\{X\}_{sol} = [k]\{u\}_{sol} = \begin{Bmatrix} 0 \\ -1000 \\ 0\cdot 5 \\ 0\cdot 5 \\ 0\cdot 5 \\ -0\cdot 5 \end{Bmatrix}$$

32

The residual force vector is defined as the original force vector minus that obtained from the displacement solution. In this case:

$$\{R\} = \{X\} - \{X\}_{sol} = \begin{Bmatrix} 0 \\ 0 \\ -0 \cdot 5 \\ -0 \cdot 5 \\ -0 \cdot 5 \\ 0 \cdot 5 \end{Bmatrix}.$$

Using the infinity norm,

$$\|X\|_\infty = \max|X_i| = 1000 \quad ; \quad \|R\|_\infty = \max|R_i| = 0 \cdot 5 \quad : \quad \frac{\|R\|_\infty}{\|X\|_\infty} = \frac{0 \cdot 5}{1000} = 0 \cdot 0005.$$

Using the Euclidian vector norm,

$$\|X\|_2 = \sqrt{\sum_i X_i^2} = 1000 \quad ; \quad \|R\|_2 = \sqrt{\sum_i R_i^2} = 1 \cdot 0 \quad : \quad \frac{\|R\|_2}{\|X\|_2} = \frac{1 \cdot 0}{1000} = 0 \cdot 001.$$

Using either vector norm the residual force vector is small relative to the original force vector.

2.6.2.2 System exhibiting ill-conditioning

Consider a system giving rise to the following set of equations:

$$\begin{Bmatrix} -1.59 \\ 1 \\ 1 \\ -1.64 \end{Bmatrix} = \begin{bmatrix} 4.855 & -4 & 1 & 0 \\ -4 & 5.855 & -4 & 1 \\ 1 & -4 & 5.855 & -4 \\ 0 & 1 & -4 & 4.855 \end{bmatrix} \begin{Bmatrix} u_1 \\ u_2 \\ u_3 \\ u_4 \end{Bmatrix}$$

Using Gaussian elimination and working to six significant figures, the upper triangular form is:

$$\begin{Bmatrix} -1.59000 \\ -0.309980 \\ 0.942827 \\ 0.004390 \end{Bmatrix} = \begin{bmatrix} 4.85500 & -4 & 1 & 0 \\ 0 & 2.55944 & -3.17610 & 1 \\ 0 & 0 & 1.70771 & -2.75907 \\ 0 & 0 & 0 & 0.006600 \end{bmatrix} \begin{Bmatrix} u_1 \\ u_2 \\ u_3 \\ u_4 \end{Bmatrix}$$

Hence the condition number for this stiffness matrix is $\dfrac{4.855}{0.0066} = 735.6$.

Performing the back-substitution step:

$$\{u\}_{sol} = \begin{Bmatrix} u_1 \\ u_2 \\ u_3 \\ u_4 \end{Bmatrix} = \begin{Bmatrix} 0.686706 \\ 1.63768 \\ 1.62674 \\ 0.665151 \end{Bmatrix}$$

Now calculate the solution force vector:

$$\{X\}_{sol} = [k]\{u\}_{sol} = \begin{bmatrix} 4.855 & -4 & 1 & 0 \\ -4 & 5.855 & -4 & 1 \\ 1 & -4 & 5.855 & -4 \\ 0 & 1 & -4 & 4.855 \end{bmatrix} \begin{bmatrix} 0.686706 \\ 1.63768 \\ 1.62674 \\ 0.665151 \end{bmatrix} = \begin{Bmatrix} -1.59002237 \\ 0.99998340 \\ 0.99994470 \\ -1.639971895 \end{Bmatrix}$$

Hence, the residual force vector $\{R\}$ is given by:

$$\{R\} = \{X\} - \{X\}_{sol} = \begin{Bmatrix} 0.00002237 \\ 0.00001660 \\ 0.00005530 \\ -0.000028105 \end{Bmatrix}$$

Using the infinity norm:

$$\frac{\|R\|_\infty}{\|X\|_\infty} = \frac{0 \cdot 00005530}{1.64} = 0 \cdot 0000337$$

This represents an error in the forces of 0.003%.

The residual displacement vector, calculated from $\{r\} = [k]^{-1}\{R\}$, is:

$$\{r\} = \begin{Bmatrix} 0.01702 \\ 0.02756 \\ 0.02754 \\ 0.01701 \end{Bmatrix}$$

It can be shown that the exact solution is:

$$\{u\}_{exact} = \begin{Bmatrix} u_1 \\ u_2 \\ u_3 \\ u_4 \end{Bmatrix} = \begin{Bmatrix} 0.7037247 \\ 1.6652256 \\ 1.6542831 \\ 0.6821567 \end{Bmatrix}$$

Hence, using the infinity norm again:

$$\frac{\|r\|_\infty}{\|u_{exact}\|_\infty} = \frac{0 \cdot 02756}{1.6652256} = 0 \cdot 0166$$

Which represents an error of 1.66%.

So, the large(ish) condition number of 735.6 has caused a displacement solution error over 550 times that of the force solution error.

Although computers can carry more decimal places than our hand calculators, think of the implications when cond(k) is greater than 10^6.

2.6.3 Sources of ill-conditioning

Ill-conditioning occurs when there is a large difference in the terms in the stiffness matrix. For simple systems of bar elements such as those considered in this section, this might arise when elements of widely disparate stiffness are used in the analysis. This is easy to spot when the element areas vary significantly, such as in structures containing bars and wires, but it should be recognised that it can also happen when the lengths of the elements vary by several orders of magnitude, and this is sometimes overlooked.

Example of Ill-conditioning

Consider a simple bar structure with 3 elements:

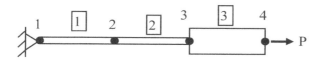

For Elements 1 & 2: $\dfrac{AE}{L} = 1$

For Element 3: $\dfrac{AE}{L} = 100$

The full set of equations is:

$$\begin{Bmatrix} X_1 \\ X_2 \\ X_3 \\ X_4 \end{Bmatrix} = \begin{bmatrix} 1 & -1 & 0 & 0 \\ -1 & 2 & -1 & 0 \\ 0 & -1 & 101 & -100 \\ 0 & 0 & -100 & 100 \end{bmatrix} \begin{Bmatrix} u_1 \\ u_2 \\ u_3 \\ u_4 \end{Bmatrix}$$

So, the set of equations to solve is:

$$\begin{Bmatrix} 0 \\ 0 \\ P \end{Bmatrix} = \begin{bmatrix} 2 & -1 & 0 \\ -1 & 101 & -100 \\ 0 & -100 & 100 \end{bmatrix} \begin{Bmatrix} u_2 \\ u_3 \\ u_4 \end{Bmatrix}$$

Working to a precision of 4 significant figures:

$$\begin{Bmatrix} 0 \\ 0 \\ P \end{Bmatrix} = \begin{bmatrix} 2 & -1 & 0 \\ 0 & 100.5 & -100 \\ 0 & -100 & 100 \end{bmatrix} \begin{Bmatrix} u_2 \\ u_3 \\ u_4 \end{Bmatrix}$$

$$\begin{Bmatrix} 0 \\ 0 \\ P \end{Bmatrix} = \begin{bmatrix} 2 & -1 & 0 \\ 0 & 100.5 & -100 \\ 0 & 0 & 0.5000 \end{bmatrix} \begin{Bmatrix} u_2 \\ u_3 \\ u_4 \end{Bmatrix}$$

$$100 - \dfrac{100}{100.5} \times 100$$
$$= 100 - 99.50(25)$$
$$= 0.5000$$

Hence, $u_4 = 2.000\,P$

The exact answer is $u_4 = 2.010\,P$, so an error of 0.5% has been introduced by the finite precision calculation.

Now repeat the calculation with $\dfrac{AE}{L} = 1000$ for Element 3:

$$\begin{Bmatrix} 0 \\ 0 \\ P \end{Bmatrix} = \begin{bmatrix} 2 & -1 & 0 \\ -1 & 1001 & -1000 \\ 0 & -1000 & 1000 \end{bmatrix} \begin{Bmatrix} u_2 \\ u_3 \\ u_4 \end{Bmatrix}$$

$$\begin{Bmatrix} 0 \\ 0 \\ P \end{Bmatrix} = \begin{bmatrix} 2 & -1 & 0 \\ 0 & 1000 & -1000 \\ 0 & -1000 & 1000 \end{bmatrix} \begin{Bmatrix} u_2 \\ u_3 \\ u_4 \end{Bmatrix}$$

$$\begin{Bmatrix} 0 \\ 0 \\ P \end{Bmatrix} = \begin{bmatrix} 2 & -1 & 0 \\ 0 & 1000 & -1000 \\ 0 & 0 & 0 \end{bmatrix} \begin{Bmatrix} u_2 \\ u_3 \\ u_4 \end{Bmatrix}$$

There is now no solution for u_4 and the problem is much the same as the case where rigid body motions are present.

- The finite element analysis system will report that the analysis is terminating due to a zero pivot.
- In some circumstances, due to rounding errors, it is possible for the pivot to become negative, which will also terminate the analysis.
- The moral of the story is that apparently normal structures can give rise to numerical problems caused by computer limitations.
- Attention must be paid to any diagnostic messages issued by the analysis package that relate to stiffness matrix conditioning.

What measures can be taken to alleviate the problem of an ill-conditioned matrix?

Sometimes processing the stiffest part of the structure first can improve the solution. Taking the same example as before, but re-numbering the elements:

Now the full set of equations is:

$$\begin{Bmatrix} 0 \\ P \\ X_3 \\ 0 \end{Bmatrix} = \begin{bmatrix} 1001 & -1000 & 0 & -1 \\ -1000 & 1000 & 0 & 0 \\ 0 & 0 & 1 & -1 \\ -1 & 0 & -1 & 2 \end{bmatrix} \begin{bmatrix} u_1 \\ u_2 \\ u_3 \\ u_4 \end{bmatrix}$$

And the equations to solve are:

$$\begin{Bmatrix} 0 \\ P \\ 0 \end{Bmatrix} = \begin{bmatrix} 1001 & -1000 & -1 \\ -1000 & 1000 & 0 \\ -1 & 0 & 2 \end{bmatrix} \begin{bmatrix} u_1 \\ u_2 \\ u_4 \end{bmatrix}$$

Working to 4 significant figures as before:

$$\begin{Bmatrix} 0 \\ P \\ 0 \end{Bmatrix} = \begin{bmatrix} 1001 & -1000 & -1 \\ 0 & 1.000 & -0.9990 \\ 0 & -0.9990 & 1.999 \end{bmatrix} \begin{bmatrix} u_1 \\ u_2 \\ u_4 \end{bmatrix}$$

$$\begin{Bmatrix} 0 \\ P \\ 0.999P \end{Bmatrix} = \begin{bmatrix} 1001 & -1000 & -1 \\ 0 & 1.000 & -0.9990 \\ 0 & 0 & 1.001 \end{bmatrix} \begin{bmatrix} u_1 \\ u_2 \\ u_4 \end{bmatrix}$$

So

$$u_4 = \frac{0.999P}{1.001} = 0.998P$$

$$u_2 - 0.999u_4 = P$$

$$u_2 = P(1 + 0.999 \times 0.998) = 1.997\,P$$

The exact answer is $u_2 = 2.001\,P$, so the error is 0.2%.

A reasonable solution has been obtained by simply re-numbering the nodes so that the stiffest terms are accumulated at the top of the matrix.

- In 1D, ill-conditioning is easy to spot when element areas vary significantly, e.g., in structures containing bars and wires.

- But can also happen when the lengths of the elements vary by orders of magnitude (AE/L).
- The most common source of ill-conditioning is failure to suppress all rigid body modes of a structure. This is discussed in more detail in Section 2.7.
- Complex structures modelled using beam and shell elements are most likely to have ill-conditioning.

A common source of ill-conditioning in these simple systems is the failure to suppress a rigid body mode of the structure. This is discussed in more detail in Section 2.7.

For more complex structures, and particularly those modelled using beam and shell elements, the problem of ill-conditioning is much more likely to arise. It can also be much more difficult to deal with.

2.7 Singular Equations: Rigid Body Modes and Mechanisms

The solution strategy suggested in Section 2.4 and illustrated by numerical examples in Section 2.5 can be applied to any structural system, provided that appropriate element stiffness matrices can be written down. No particular emphasis was placed on step (iii) of the suggested strategy (the removal of the equations associated with the prescribed (zero) displacements) but in practice many errors arise from an inadequate definition of the supports of a structure.

Consider the simple triangular framework illustrated below. For each member AE/L = 2.

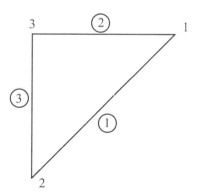

Figure 2.6 Simple triangular framework.

39

The stiffness matrix of each element can be written down as in previous examples. The assembled global stiffness matrix is:

$$[k] = \begin{bmatrix} 3 & 1 & -1 & -1 & -2 & 0 \\ 1 & 1 & -1 & -1 & 0 & 0 \\ -1 & -1 & 1 & 1 & 0 & 0 \\ -1 & -1 & 1 & 3 & 0 & -2 \\ -2 & 0 & 0 & 0 & 2 & 0 \\ 0 & 0 & 0 & -2 & 0 & 2 \end{bmatrix}$$

Now suppose that the framework is subjected to an arbitrary self-equilibrating set of forces. It might be supposed that the displacements of the nodes could be computed in terms of the applied forces. Following the methodology of the worked example,

$$\begin{bmatrix} X_1 \\ Y_1 \\ X_2 \\ Y_2 \\ X_3 \\ Y_3 \end{bmatrix} = \begin{bmatrix} 3 & 1 & -1 & -1 & -2 & 0 \\ 1 & 1 & -1 & -1 & 0 & 0 \\ -1 & -1 & 1 & 1 & 0 & 0 \\ -1 & -1 & 1 & 3 & 0 & -2 \\ -2 & 0 & 0 & 0 & 2 & 0 \\ 0 & 0 & 0 & -2 & 0 & 2 \end{bmatrix} \begin{bmatrix} u_1 \\ v_1 \\ u_2 \\ v_2 \\ u_3 \\ v_3 \end{bmatrix}$$

Eliminating entries below the leading diagonal in the first column:

$$\begin{bmatrix} X_1 \\ Y_1 - \tfrac{1}{3}X_1 \\ X_2 + \tfrac{1}{3}X_1 \\ Y_2 + \tfrac{1}{3}X_1 \\ X_3 + \tfrac{2}{3}X_1 \\ Y_3 \end{bmatrix} = \begin{bmatrix} 3 & 1 & -1 & -1 & -2 & 0 \\ 0 & \tfrac{2}{3} & -\tfrac{2}{3} & -\tfrac{2}{3} & \tfrac{2}{3} & 0 \\ 0 & -\tfrac{2}{3} & \tfrac{2}{3} & \tfrac{2}{3} & -\tfrac{2}{3} & 0 \\ 0 & -\tfrac{2}{3} & \tfrac{2}{3} & \tfrac{8}{3} & -\tfrac{2}{3} & -2 \\ 0 & \tfrac{2}{3} & -\tfrac{2}{3} & -\tfrac{2}{3} & \tfrac{2}{3} & 0 \\ 0 & 0 & 0 & -2 & 0 & 2 \end{bmatrix} \begin{bmatrix} u_1 \\ v_1 \\ u_2 \\ v_2 \\ u_3 \\ v_3 \end{bmatrix}$$

Eliminating entries below the leading diagonal in the second column:

$$
\begin{bmatrix}
X_1 \\
Y_1 - \tfrac{1}{3}X_1 \\
X_2 + Y_1 \\
Y_2 + Y_1 \\
X_3 + X_1 - Y_1 \\
Y_3
\end{bmatrix}
=
\begin{bmatrix}
3 & 1 & -1 & -1 & -2 & 0 \\
0 & \tfrac{2}{3} & -\tfrac{2}{3} & -\tfrac{2}{3} & \tfrac{2}{3} & 0 \\
0 & 0 & 0 & 0 & 0 & 0 \\
0 & 0 & 0 & 2 & 0 & -2 \\
0 & 0 & 0 & 0 & 0 & 0 \\
0 & 0 & 0 & -2 & 0 & 2
\end{bmatrix}
\begin{bmatrix}
u_1 \\
v_1 \\
u_2 \\
v_2 \\
u_3 \\
v_3
\end{bmatrix}
$$

The entries below the leading diagonal in the third column are already zero. Eliminating entries below the leading diagonal in the fourth column:

$$
\begin{bmatrix}
X_1 \\
Y_1 - \tfrac{1}{3}X_1 \\
X_2 + Y_1 \\
Y_2 + Y_1 \\
X_3 + X_1 - Y_1 \\
Y_3 + Y_2 + Y_1
\end{bmatrix}
=
\begin{bmatrix}
3 & 1 & -1 & -1 & -2 & 0 \\
0 & \tfrac{2}{3} & -\tfrac{2}{3} & -\tfrac{2}{3} & \tfrac{2}{3} & 0 \\
0 & 0 & 0 & 0 & 0 & 0 \\
0 & 0 & 0 & 2 & 0 & -2 \\
0 & 0 & 0 & 0 & 0 & 0 \\
0 & 0 & 0 & 0 & 0 & 0
\end{bmatrix}
\begin{bmatrix}
u_1 \\
v_1 \\
u_2 \\
v_2 \\
u_3 \\
v_3
\end{bmatrix}
$$

This is the system of equations to be solved by back-substitution. There is clearly a problem because three of the rows of the matrix to be inverted contain only zeros, and there are only three useful equations in the six unknown displacements. The global stiffness matrix [k] has a rank deficiency of three.

It is interesting to investigate the three equations that are implied by the deficiencies in the matrix. The first equation yielded by a row of null entries is $X_2 + Y_1 = 0$. Inspecting the structure, this is an equation of static equilibrium representing the taking of moments about node 3. The second such equation is $X_3 + X_1 - Y_1 = 0$, representing the taking of moments about node 2. The third equation is $Y_3 + Y_2 + Y_1 = 0$, representing the equilibrium of forces in the global Y direction. Since it was stated at the beginning of the problem that the forces on the frame were in equilibrium, these equations have yielded no additional information.

There is a rank deficiency of three in the global stiffness matrix, this is because there are three possible rigid body modes of the frame. It can translate along the global x axis or along the global y axis, and it can spin about the origin. All possible rigid body motions of the frame can be expressed as some combination of these three modes. If the equations to be solved in the stiffness method approach are not to be singular, it is necessary to suppress sufficiently many degrees of freedom to prevent these rigid body modes. In the case of the simple triangular frame, it is easy to choose a combination of freedoms to suppress that will achieve this requirement. It is useful to work in a logical manner, suppressing one freedom at a time and considering what rigid body modes remain. One possible set of constraints is as follows:

i) $u_2 = 0$. This prevents translation along global x.

+ ii) $v_2 = 0$. The structure can now only rotate about node 2.

+ iii) $u_3 = 0$. No rigid body modes remain.

Often the appropriate constraints will be suggested by the manner in which the real structure is supported. For example, the above set of constraints would be most appropriate for a structure pinned to a rigid support at node 2 and with a roller-guide running parallel to the Y axis at node 3.

The situation can be much more difficult when complicated frameworks, particularly in three dimensions, are designed. In principle, the choice of an appropriate set of six degrees of freedom will provide adequate support for a

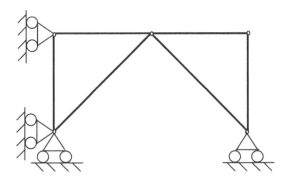

Figure 2.7 Framework with rigid body motions.

structure in three dimensions, as a set of three did in two dimensions. Problems arise when the structure contains internal mechanisms, when additional suppressions will be required. This is illustrated in the diagram below, in which four suppressions are required because the structure itself has one mechanism.

It has been demonstrated that the finite element form of the equilibrium equation cannot be solved when rigid body motions are present. In practice, it is likely that small (spurious) stiffness terms will arise in place of the zero values, perhaps because the members are not aligned with global axes and transformations produce rounding errors. In this case, the finite element computations might produce a solution, but it will have no true meaning. The problem is usually extremely ill-conditioned because the difference between the true stiffness terms and the spurious terms is often very large. Hence, a high condition number for a stiffness matrix might be indicative that the structure contains unrestrained rigid body modes.

2.8 Symmetry, Antisymmetry and Asymmetry

It is apparent that the number of equations to be solved will grow rapidly as additional nodes and elements are introduced. There are several techniques that can be used to reduce the amount of computation that is required for a solution, and one of the most powerful of these is based on the notions of symmetry and antisymmetry. Analysis procedures for symmetrical two-dimensional frameworks subjected to various loading conditions are discussed in Sections 2.8.1 to 2.8.3. Note that the boundary condition specified on the line of geometrical symmetry applies explicitly to the two-dimensional system with two translational degrees of freedom at each node. Additional constraints will be required for three dimensional systems and for systems featuring elements such as beams and shells with rotational freedoms. Note further that large displacement effects would often render the antisymmetrical boundary condition inappropriate for non-linear analyses, and that the procedure for asymmetrical loading depends upon the linear combination of loading cases and is therefore, by definition, inappropriate for non-linear systems.

2.8.1 Symmetrical loading

Consider the two-dimensional pin-jointed framework illustrated below.

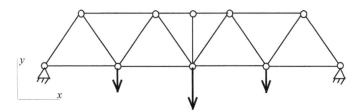

Figure 2.8 Two-dimensional pin-jointed framework.

There are ten nodes, of which two are fixed in space. There are two degrees of freedom per node, and therefore sixteen simultaneous equations will remain after elimination of the suppressed freedoms. The number of equations can be reduced by a factor of two by noting the line of symmetry of geometry and of loading about the middle of the structure in the *x* direction. The two nodes on the symmetry line cannot move in the *x* direction, because if they did symmetry would be violated. It is, therefore, possible to model one half of the structure, with a symmetry constraint on the symmetry plane.

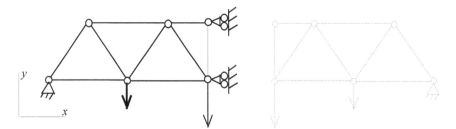

Figure 2.9 Pin-jointed framework with symmetry.

The half-structure has six nodes and, hence, twelve degrees of freedom. Four of these are eliminated by the boundary conditions, and, therefore, eight simultaneous equations remain to be solved. Note that the symmetry constraint implies that a mirror image of the structure and of the loading lies on the other side of the symmetry boundary. For this reason, only one half of the load on the symmetry boundary should be applied to the half-structure and only one half of the stiffness of the member on the symmetry plane should be included in the global stiffness matrix of the half-structure.

2.8.2 Antisymmetrical loading

Sometimes the geometry of a structure is symmetrical about a plane, but the loading is antisymmetrical. In this case the structure can still be split at the plane of geometrical symmetry, but now antisymmetrical boundary conditions are applied. Consider the same structure as that illustrated above, but with antisymmetrical loading.

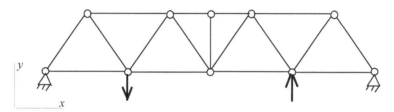

Figure 2.10 Pin-jointed framework with antisymmetrical loading.

Now the nodes on the symmetry line can move in the *x* direction but, because of the symmetry of the structure, the action of the forces on the left of the line, pushing the nodes downwards, is perfectly balanced by the forces on the right pushing the nodes upwards. Hence, the nodes do not move in the vertical direction, along the line of symmetry. Again, one half of the structure is modelled and the constraints indicated below are applied.

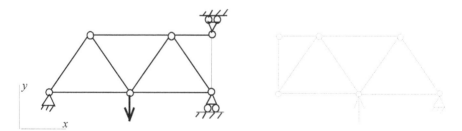

Figure 2.11 Pin-jointed framework with antisymmetry.

45

2.8.3 *Asymmetrical loading*

For symmetrical and antisymmetrical loadings, as described in the preceding two sections, one half of the structure is modelled. This results in a very substantial reduction in the computer resources required for the solution. In many cases, the loading is neither symmetrical nor antisymmetrical. In this case, two analyses can be performed, one using symmetrical boundary conditions and one using antisymmetrical boundary conditions. The results can be added together to give the true loading condition. This is illustrated diagrammatically below.

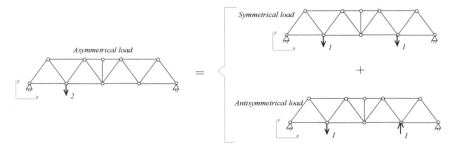

Figure 2.12 Pin-jointed framework with asymmetrical loading.

The two loading conditions on the right can be analysed using the half-structures with the appropriate boundary conditions, and then the results can be added together to give the true loading on the structure. Whether this procedure will result in a worthwhile saving in computer resources depends on the type of structure that is being analysed. Although this procedure has been very common in the past, when computer resources were more limited, the extra man-time that is required to create a model, to post-process the results and to assess the structure often dictates against its use in the modern analysis environment.

2.9 Thermal Loads

In many applications the engineer will be concerned with structures that are stressed by thermal loads as well as by mechanical loads and this in itself is sufficient reason to introduce temperature into the finite element formulation. However, the application of thermal loads to structural models proves to be a very powerful tool for the study of systems in which there are no real temperature gradients. Consider, for example, a system in which a bolt is pre-stressed before further mechanical loads are applied. The simplest way to model such a system is to set up the geometry with the bolt unloaded and then to reduce the temperature of the bolt so that it tends to get shorter. This shortening will be restrained by the remainder of the structure, thus setting up a pre-stress situation. This methodology is particularly useful when the bolt will be pre-stressed by a turning it (or a nut) through a given number of turns on a thread of known pitch. Thermal loads can be used in a similar manner to model interference fits.

2.9.1 Thermal loads in a bar element

Figure 2.13 One dimensional element with thermal loads.

Consider once again the simple bar element with nodes *i* and *j*. If the bar is loaded by the mechanical forces shown and increased in temperature by an amount ΔT, the total strain is given by:

$$\varepsilon_x = \frac{\partial \bar{u}}{\partial x} = \frac{\sigma_x}{E} + \alpha \, \Delta T \, .$$

$$\uparrow \qquad \uparrow$$

Mechanical   Thermal

Strain.   Strain.

From equilibrium:

$$F = -\bar{X}_i = \bar{X}_j \qquad\qquad \sigma_x = \frac{-\bar{X}_i}{A}$$

$$\int_{\bar{u}_i}^{\bar{u}_j} d\bar{u} = \int_{x_i}^{x_j} \left(\frac{\sigma_x}{E} + \alpha . \Delta T \right) . dx$$

$$\bar{u}_j - \bar{u}_i = \left(-\frac{\bar{X}_i}{AE} + \alpha . \Delta T \right) . L$$

Hence,

$$\overline{X}_i - AE\alpha\Delta T = \frac{AE}{L}\left(\overline{u}_i - \overline{u}_j\right)$$

And similarly:

$$\overline{X}_j + AE\alpha\Delta T = \frac{AE}{L}\left(-\overline{u}_i + \overline{u}_j\right)$$

In matrix form:

$$\left\{\begin{matrix} \overline{X}_i \\ \overline{X}_j \end{matrix}\right\} + AE\alpha\Delta T \left\{\begin{matrix} -1 \\ 1 \end{matrix}\right\} = \frac{AE}{L}\begin{bmatrix} 1 & -1 \\ -1 & 1 \end{bmatrix}\left\{\begin{matrix} \overline{u}_i \\ \overline{u}_j \end{matrix}\right\}$$

   ↑   ↑

Nodal   Thermal

forces.   'forces'.

2.9.2 Numerical example of pre-stressing of bolt using thermal loads

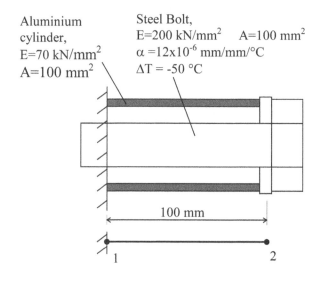

Element 1: Aluminium Cylinder

$$\left\{ \begin{matrix} \bar{X}_1 \\ \bar{X}_2 \end{matrix} \right\} = \frac{A_1 E_1}{L} \begin{bmatrix} 1 & -1 \\ -1 & 1 \end{bmatrix} \left\{ \begin{matrix} \bar{u}_1 \\ \bar{u}_1 \end{matrix} \right\}$$

Element 2: Steel Bolt

$$\left\{ \begin{matrix} \bar{X}_1 \\ \bar{X}_2 \end{matrix} \right\} + A_2 E_2 \alpha \Delta T \left\{ \begin{matrix} -1 \\ 1 \end{matrix} \right\} = \frac{A_2 E_2}{L} \begin{bmatrix} 1 & -1 \\ -1 & 1 \end{bmatrix} \left\{ \begin{matrix} \bar{u}_1 \\ \bar{u}_2 \end{matrix} \right\}$$

Assembling:

$$\left\{ \begin{matrix} X_1 - A_2 E_2 \alpha \Delta T \\ X_2 + A_2 E_2 \alpha \Delta T \end{matrix} \right\} = \frac{A_1 E_1 + A_2 E_2}{L} \begin{bmatrix} 1 & -1 \\ -1 & 1 \end{bmatrix} \left\{ \begin{matrix} u_1 \\ u_2 \end{matrix} \right\}$$

Taking note of the boundary conditions:

$$\left\{ \begin{matrix} X_1 - A_2 E_2 \alpha \Delta T \\ A_2 E_2 \alpha \Delta T \end{matrix} \right\} = \frac{A_1 E_1 + A_2 E_2}{L} \begin{bmatrix} 1 & -1 \\ -1 & 1 \end{bmatrix} \left\{ \begin{matrix} 0 \\ u_2 \end{matrix} \right\}$$

Substituting in the values:

$$\left\{ \begin{matrix} X_1 + 12000 \\ -12000 \end{matrix} \right\} = \frac{270 \times 10^5}{100} \begin{bmatrix} 1 & -1 \\ -1 & 1 \end{bmatrix} \left\{ \begin{matrix} 0 \\ u_2 \end{matrix} \right\}$$

Solving for u_2:

$$-12000 = 270000(-1.0 + 1.u_2)$$

$$\therefore u_2 = \frac{-12}{270} = -0.0444 \text{ mm}$$

Subst. back to get X_1:

$$X_1 + 12000 = 270000\{1 \times 0 + (-1) \times (-0.0444)\} = 12000$$

$$\therefore X_1 = 0$$

Now calculate the *mechanical* strain in each element:

$$\frac{\sigma_x}{E} = \frac{\partial \bar{u}}{\partial x} - \alpha \Delta T$$

Steel:

$$\frac{\sigma_x}{E} = \frac{-0.0444 - 0}{100} - 12 \times 10^{-6}.(-50) = 156 \times 10^{-6}$$

So

$$\sigma_{steel} = 200 \times 10^3 \times 156 \times 10^{-6} = 31.2 \text{ N/mm}^2$$

Aluminium:

$$\frac{\sigma_x}{E} = \frac{-0.0444 - 0}{100} = -444 \times 10^{-6}$$

So

$$\sigma_{aluminium} = 70 \times 10^3 \times (-444 \times 10^{-6}) = -31.2 \text{ N/mm}^2$$

Applications

1) Bolts tightened by a known number of turns.
2) Mechanical misfits.
3) Deformation of components during machining.

Chapter 3

The Finite Element Formulation
One-Dimensional Problems

The analysis presented in Section 2.1 resulted in an expression for the stiffness matrix of a bar in its local co-ordinate system. The derivation was based on an immediate appeal to equilibrium to establish the force and, hence, stress distribution in the bar. This method works well for the bar element of constant cross-section, for which exact expressions for the entries in the stiffness matrix are readily deduced. This bar element might be regarded as unusual in that the finite element formulation is exact. In more general cases, the development of the element stiffness matrix might not be so straightforward, and some approximation might be necessary. In these situations there are alternative approaches that might be easier to apply to generate the stiffness terms. An alternative approach to the formulation of the one-dimensional problem will now be presented.

3.1 The Fundamental Equations

3.1.1 The equilibrium equation

The finite element method represents a numerical procedure for the solution of the fundamental or governing equations of a physical system. For problems in one-dimensional elasticity, the fundamental equation is the equation of equilibrium. Consider the elastic bar illustrated below.

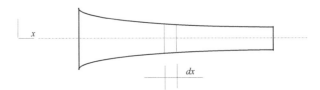

Figure 3.1 Elastic bar.

Assume that the cross-sectional area of the bar is A_x and that the tensile stress on any cross-section is σ_x.

The force acting to the left of the element dx is:

$\sigma_x \cdot A_x$.

The force acting to the right of the element is:

$(\sigma x + (d\sigma_x/dx)\,dx)\,.\,(Ax + (dA_x/dx)\,dx)$.

For equilibrium of the element, these forces must be equal:

$$\sigma_x A_x = \sigma_x A_x + \sigma_x \cdot \left(\frac{dA_x}{dx} \right) dx + A_x \left(\frac{d\sigma_x}{dx} \right) dx + \left(\frac{dA_x}{dx} \right) dx \left(\frac{d\sigma_x}{dx} \right) dx .$$

Neglecting quantities of second order:

$$\frac{d}{dx} \left(A_x\, \sigma_x \right) = 0 .$$

This is the stress form of the fundamental equation of equilibrium of an elastic bar.

An alternative expression of the same equation can be found by using Hooke's Law relating stress to strain, and then substituting the definition of strain in terms of displacement:

$$\sigma_x = E_x\, \varepsilon_x = E_x\, \frac{du}{dx} .$$

The displacement form of the equation of equilibrium of a one-dimensional elastic system is therefore:

$$\frac{d}{dx}\left(A_x \; E_x \; \frac{du}{dx}\right) = 0 .$$

(3.1)

3.1.2 The principle of virtual displacements

An alternative expression of the equilibrium equation is provided by the principle of virtual displacements. One expression of this principle is that

> 'when a small virtual displacement is imposed on a body that is in equilibrium, the additional strain energy generated in the body is equal to the virtual work done by the external forces acting on the body.'

The mathematical expression of the principle is:

$$\int_{Vol} \{\delta \; \varepsilon\}^T \{\sigma\} \; dVol = \int_{Vol} \{\delta \; u\}^T \{f\}_{body} \; dVol + \int_{Surf} \{\delta \; u\}^T \{f\}_{surf} \; dSurf + \{\delta \; u\}^T_{nodes} \{X\}_{nodes} .$$

(3.2)

For a system of one-dimensional elements this can be simplified to:

$$\sum_{elems} \int_{x_i}^{x_j} \delta \; \varepsilon \cdot \sigma \; A_x \; dx = \sum_{elems} \int_{x_i}^{x_j} \delta \; u \cdot W \; A_x \; dx + \{\delta \; u\}^T \{X\}$$

(3.3)

where W is the body force per unit volume.

3.2 The Shape Function

It was stated in the introduction, Chapter 1, that the finite element method is based upon the assumption that a structure can be broken down into elements, the behaviour of each of which is understood or can be postulated. The simple bar element developed in Chapter 2 is an example of the former. Consider now an alternative approach, in which the distribution of a fundamental variable in an element is simply assumed. In a structural element this variable might be stress, strain or displacement, or a combination of the three. In other field problems alternative variables such as temperature (heat transfer) or a potential function (one-dimensional fluid flow) might be chosen. Throughout this chapter, attention will be focused on the *displacement method*, in which the fundamental variable is taken to be the distribution of displacement throughout an element (and a structure).

Consider again the simple bar element with nodes i and j. The forces at nodes i and j are X_i and X_j, and the displacements are u_i and u_j.

Figure 3.2 One dimensional bar element.

Assume that the displacement at all points in the element can be expressed by a function $u(x)$.

The function must satisfy the boundary conditions, $u = u_i$ at node i and $u = u_j$ at node j. This requirement will be satisfied if a function of the form

$$u = N_i u_i + N_j u_j \tag{3.4}$$

is chosen, in which N_i adopts a value of zero at $x = x_j$, and of one at $x = x_i$, and a similar function is chosen for N_j.

The functions N_i, N_j are referred to as the *shape functions* of the element. There are several requirements that the shape functions must satisfy if an element is to perform satisfactorily, and these will be discussed later. In many cases, the shape functions that are most intuitively obvious perform very well in an element formulation, but in others, relatively obscure functions with particular mathematical properties are more appropriate. The shape function is an inherent attribute of a finite element, and, as such, is not usually under direct user control. Most commercial finite element systems will offer many types of element, each based upon its own particular shape functions, and the user will apply simple heuristics to select the most appropriate element type for a given application. Often the element shape functions will not be explicitly declared by the system vendor.

For the bar element, the most obvious choices of shape function are those that imply a linear variation of the displacement between the nodes.

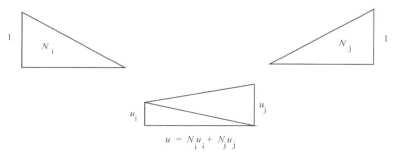

Figure 3.3 Linear shape functions.

The shape functions illustrated above are:

$$N_i = \frac{x_j - x}{L} \qquad : \qquad N_j = \frac{x - x_i}{L} \tag{3.5}$$

These shape functions define one possible formulation for a bar element. The derivatives of the shape functions prescribe the variation of displacement (*du/dx*) inside an element. The linear shape functions illustrated above imply constant strain in an element, and, therefore, the equilibrium Equation (3.1) will be violated inside the element unless the term $A_x E_x$ is constant. Since this is not true in the general case, a finite element analysis based on the linear shape function cannot give precisely accurate results for the distribution of strain and stress in an isotropic bar of varying cross-sectional area.

3.3 The Finite Element Equations

3.3.1 Algebraic form

For the bar element with the above shape functions,

$$u = N_i \, u_i + N_j \, u_j = \frac{x_j - x}{L} . u_i + \frac{x - x_i}{L} . u_j . \tag{3.6}$$

The strain in the bar is:

$$\varepsilon = \frac{du}{dx} = -\frac{\left(u_i - u_j\right)}{L} . \tag{3.7}$$

The stress, from Hooke's Law, is:

$$\sigma = E\varepsilon = -\frac{E\left(u_i - u_j\right)}{L} . \tag{3.8}$$

Now equilibrium of the system will be maintained if Equation (3.3) is satisfied, but the evaluation of the terms in this equation requires expressions for the distribution of displacement, stress and strain within the elements. The basis of the finite element method is that the true distribution of these variables can be approximated by equations like the three listed above, based on the assumed element shape functions. Substituting Equations (3.6), (3.7), and (3.8) into Equation (3.3):

$$\sum_{elems} \int_{x_i}^{x_j} \left(-\frac{\left(\delta u_i - \delta u_j\right)}{L}\right) \cdot \left(-\frac{E\left(u_i - u_j\right)}{L}\right) A_x \, dx = \sum_{elems} \int_{x_i}^{x_j} \left(\frac{\left(x_j - x\right)}{L} \delta u_i + \frac{\left(x - x_i\right)}{L} \delta u_j\right) \cdot W \, A_x \, dx + \sum_{nodes} \{\delta u_i \, X_i\}$$

$$\sum_{elems} \frac{E\left(\delta u_i - \delta u_j\right)\left(u_i - u_j\right)}{L^2} \int_{x_i}^{x_j} A_x \, dx = \sum_{elems} \int_{x_i}^{x_j} \left(\frac{\left(x_j - x\right)}{L} \delta u_i + \frac{\left(x - x_i\right)}{L} \delta u_j\right) W \, A_x \, dx + \sum_{nodes} \delta u_i \, X_i$$

$$\tag{3.9}$$

Since this equation holds for all arbitrary values of the virtual displacement, the coefficients of each virtual displacement term must be the same on both sides of the equation. Equating the coefficients of each of the virtual displacements, δu_i, yields one equation relating the force in the direction of that displacement to the displacements of the structure. There are, therefore, as many equations as there are degrees of freedom. The equation featuring the force at the *i*th node is presented below.

$$\sum_{elems} \frac{E\left(u_i - u_j\right)}{L^2} \int_{x_i}^{x_j} A_x \, dx = \sum_{elems} \int_{x_i}^{x_j} \frac{\left(x_j - x\right)}{L} W \, A_x \, dx + X_i \qquad (3.10)$$

3.3.2 Matrix form

The equations developed in the preceding section are most often expressed in matrix form. Equation (3.6) becomes:

$$u = N_i \, u_i + N_j \, u_j = \begin{bmatrix} N_i & N_j \end{bmatrix} \begin{bmatrix} u_i \\ u_j \end{bmatrix} = [N]\{u\} \, . \qquad (3.11)$$

The strain in the bar is:

$$\varepsilon = \frac{du}{dx} = \begin{bmatrix} \dfrac{dN_i}{dx} & \dfrac{dN_j}{dx} \end{bmatrix} \begin{bmatrix} u_i \\ u_j \end{bmatrix} = [B]\{u\} \, . \qquad (3.12)$$

where [B] is a matrix of derivatives of the shape functions.

The stress, from Hooke's Law, is:

$$\sigma = E\varepsilon = E[B]\{u\}. \qquad (3.13)$$

Substituting Equations (3.11), (3.12), and (3.13) into Equation (3.3):

$$\sum_{elems} \int_{x_i}^{x_j} \{\delta \, u\}^T [B]^T E[B]\{u\} \, A_x \, dx = \sum_{elems} \int_{x_i}^{x_j} \{\delta \, u\}^T [N]^T \cdot W \, A_x \, dx + \{\delta \, u\}^T \{X\}$$

The vectors of displacements and of virtual displacements are independent of the variable of integration and can be taken outside the integrals.

$$\sum_{elems} \{\delta \, u\}^T \int_{x_i}^{x_j} [B]^T E[B] \, A_x \, dx \, \{u\} = \sum_{elems} \{\delta \, u\}^T \int_{x_i}^{x_j} [N]^T \cdot W \, A_x \, dx + \{\delta \, u\}^T \{X\}$$

In the above equation, the terms inside the summations are in local element co-ordinates and the vectors and matrices have dimensions related to the number of degrees of freedom in the element. By expanding the vectors and matrices to include all of the freedoms in the system (by padding the [B] and [N] matrices with zeros as required) the vectors can be taken outside the summations.

$$\{\delta\,\mathrm{u}\}^T\left(\sum_{elems}\int_{x_i}^{x_j}[B]^T\,E[B]\,A_x\,dx\right)\{\mathrm{u}\} = \{\delta\,\mathrm{u}\}^T\left(\sum_{elems}\int_{x_i}^{x_j}[N]^T\cdot W\,A_x\,dx\right)+\{\delta\,\mathrm{u}\}^T\{\mathrm{X}\}$$

Since this is true for arbitrary values of the virtual displacements, the coefficients of each term of the vector must be the same on both sides of the equation. This yields as many equations as there are terms in the virtual displacement vector, and can be expressed in matrix form as follows:

$$[k]\{u\} = \{F\}. \qquad (3.14)$$

where the global stiffness matrix [k] is simply the assembly of the element stiffness matrices $[k]_{elem}$:

$$[k]_{elem} = \int_{x_i}^{x_j}[B]^T\,E[B]\,A_x\,dx\;. \qquad (3.15)$$

and the global force vector {F} is:

$$\{F\} = \left(\sum_{elems}\int_{x_i}^{x_j}[N]^T\cdot W\,A_x\,dx\right)+\{X\}\;. \qquad (3.16)$$

In the absence of body forces, W, the global force vector is just the vector of externally-applied nodal loads.

3.4 The Element Stiffness Matrix for a 2-Node Bar with Linear Shape Functions

The element stiffness matrix can be calculated directly from Equation (3.15). The two-node bar element [B] is a 1×2 matrix (Equation (3.12)) and, therefore, $[k]_{elem}$ is a 2×2 matrix. From Equations (3.5), (3.11) and (3.12),

$$[B] = \left[-\frac{1}{L}\quad\frac{1}{L}\right].$$

Then:

$$[B]^T[B] = \begin{bmatrix} -\dfrac{1}{L} \\ \dfrac{1}{L} \end{bmatrix} \begin{bmatrix} -\dfrac{1}{L} & \dfrac{1}{L} \end{bmatrix} = \dfrac{1}{L^2} \begin{bmatrix} 1 & -1 \\ -1 & 1 \end{bmatrix}.$$

Since the $[B]^T[B]$ matrix contains only constants, it can be taken outside the integral.

$$[k]_{elem} = \dfrac{1}{L^2} \begin{bmatrix} 1 & -1 \\ -1 & 1 \end{bmatrix} \int_{x_i}^{x_j} E\ A_x\ dx \tag{3.17}$$

Up to this point no explicit assumption regarding the axial variation of E and A has been made. Two particular cases will now be considered.

3.4.1 Bar of constant Young's modulus and of constant cross-section

If the Young's modulus (E) and the cross-sectional area (A) are constant, Equation (3.17) can be written:

$$[k]_{elem} = \dfrac{1}{L^2} \begin{bmatrix} 1 & -1 \\ -1 & 1 \end{bmatrix} \int_{x_i}^{x_j} E\ A\ dx = \dfrac{AE}{L} \begin{bmatrix} 1 & -1 \\ -1 & 1 \end{bmatrix}$$

$$\begin{bmatrix} X_i \\ X_j \end{bmatrix} = \dfrac{AE}{L} \begin{bmatrix} 1 & -1 \\ -1 & 1 \end{bmatrix} \begin{bmatrix} u_i \\ u_j \end{bmatrix} \tag{3.18}$$

Note that this equation is identical to Equation (2.1). In this case, the principle of virtual displacements, together with the assumed shape functions, has yielded the exact solution for the stiffness matrix, since the shape function chosen for displacement happens to be the exact solution of the displacement form of the equilibrium equation (Equation (3.1)).

3.4.2 Bar with linear taper

Consider a bar with a constant Young's modulus (E), and with cross-sectional area varying linearly from A_i to A_j.

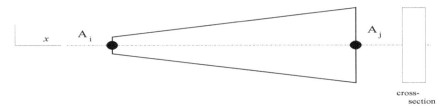

cross-
section

Figure 3.4 Bar with linear taper.

The area at any point along the bar can be written as:

$$A = \left(\frac{x_j - x}{L}\right) A_i + \left(\frac{x - x_i}{L}\right) A_j.$$

Equation (3.17) becomes:

$$[k]_{elem} = \frac{1}{L^2} \begin{bmatrix} 1 & -1 \\ -1 & 1 \end{bmatrix} \int\limits_{x_i}^{x_j} E \left(\left(\frac{x_j - x}{L}\right) A_i + \left(\frac{x - x_i}{L}\right) A_j \right) dx$$

Performing the integration and evaluating the integral,

$$[k]_{elem} = \frac{(A_i + A_j)}{2} \frac{E}{L} \begin{bmatrix} 1 & -1 \\ -1 & 1 \end{bmatrix}$$

Hence, in the absence of body forces the element load/displacement relationship is:

$$\begin{bmatrix} X_i \\ X_j \end{bmatrix} = \frac{\overline{A}E}{L} \begin{bmatrix} 1 & -1 \\ -1 & 1 \end{bmatrix} \begin{bmatrix} u_i \\ u_j \end{bmatrix} \tag{3.19}$$

where $A = (A_i + A_j)/2$ is the average of the cross-sectional areas at the two ends.

The exact expression for the stiffness matrix, obtained by integration of the equilibrium equation, is:

$$[k]_{exact} = \frac{(A_j - A_i)E}{L \ln \dfrac{A_j}{A_i}} \begin{bmatrix} 1 & -1 \\ -1 & 1 \end{bmatrix}$$

Hence, the stiffness matrix derived by the principle of virtual work, using the assumed (and now inaccurate) shape functions to evaluate the strain energy in the element, is an approximation of the true stiffness matrix of the tapered bar.

For the particular case where $A_j = 2A_i$ the approximate equation is:

$$[k] = \frac{3A_iE}{2L} \begin{bmatrix} 1 & -1 \\ -1 & 1 \end{bmatrix} = \frac{1.5A_iE}{L} \begin{bmatrix} 1 & -1 \\ -1 & 1 \end{bmatrix}$$

whilst the exact relationship is:

$$[k]_{exact} = \frac{(2A_i - A_i)E}{L \ln \dfrac{2A_i}{A_i}} \begin{bmatrix} 1 & -1 \\ -1 & 1 \end{bmatrix} = \frac{A_iE}{0.6931L} \begin{bmatrix} 1 & -1 \\ -1 & 1 \end{bmatrix}$$

$$= \frac{1.4427A_iE}{L} \begin{bmatrix} 1 & -1 \\ -1 & 1 \end{bmatrix}$$

The approximation is 4.0% higher than the exact value. The size of the error depends upon the relative sizes of A_i and A_j. For example, when $A_j = 3A_i$ the approximation is 9.9% too high.

When modelling a tapering rod with linear elements, it is necessary to use more than one element to get a good result.

Chapter 4

The Finite Element Formulation

Two-Dimensional Problems

In Chapter 3, the finite element formulation of the bar element was developed. In this chapter, a similar methodology will be followed in order to develop a finite element representation of a two-dimensional elastic continuum.

4.1 The Fundamental Equations

4.1.1 Elasticity of a continuum

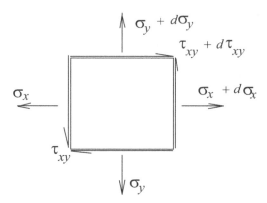

Figure 4.1 Infinitesimal rectangular element.

Consider an infinitesimal rectangular element of an elastic continuum in two dimensions. Assume that the element is of unit thickness, and that there are no body-forces acting inside the element.

The force acting to the left on the element is:

$\sigma_x \cdot d_y + \tau_{xy} \cdot dx.$

The force acting to the right on the element is:

$$\left(\sigma_x + \frac{\partial \sigma_x}{\partial x} . dx \right) . dy + \left(\tau_{xy} + \frac{\partial \tau_{xy}}{\partial y} . dy \right) . dx$$

For equilibrium of the element, these forces must be equal.

$$\frac{\partial \sigma_x}{\partial x} + \frac{\partial \tau_{xy}}{\partial y} = 0 \tag{4.1a}$$

Similarly, by consideration of the equilibrium in the y direction:

$$\frac{\partial \tau_{xy}}{\partial x} + \frac{\partial \sigma_y}{\partial y} = 0 \tag{4.1b}$$

Equations (4.1) represent the stress form of the fundamental equations of equilibrium of a 2D elastic continuum. They contain three unknowns, and a further equation is required. The definition of strains is:

$$\varepsilon_x = \frac{\partial u}{\partial x} \qquad : \qquad \varepsilon_y = \frac{\partial v}{\partial y} \qquad : \qquad \gamma_{xy} = \frac{\partial u}{\partial y} + \frac{\partial v}{\partial x}.$$

Eliminating u and v from these three equations yields the compatibility equation:

$$\frac{\partial^2 \varepsilon_x}{\partial y^2} + \frac{\partial^2 \varepsilon_y}{\partial x^2} = \frac{\partial^2 \gamma_{xy}}{\partial x \partial y} \tag{4.2}$$

63

To find the displacement forms of the equilibrium equations, Hooke's Law could be used to eliminate stresses and then the two equations could be represented in terms of displacement simply by substituting the above definitions of strain. However, it is demonstrated in Section 3.3 that the stiffness matrix of a one-dimensional element could be developed by assuming a distribution of displacement within the element and then appealing to the principle of virtual displacements as an indirect method of enforcement of equilibrium. A similar method can be applied to the development of the stiffness matrix of the continuum element.

4.1.2 *Vector notation*

It is useful to define stress and strain vectors at every point in the continuum.

$$\{\sigma\} = \begin{bmatrix} \sigma_x \\ \sigma_y \\ \tau_{xy} \end{bmatrix} \qquad \{\varepsilon\} = \begin{bmatrix} \varepsilon_x \\ \varepsilon_y \\ \gamma_{xy} \end{bmatrix}$$

Similarly, the displacement at every point is expressed as a vector.

$$\{u\} = \begin{bmatrix} u \\ v \end{bmatrix}$$

4.1.3 *The stress/strain relationships*

Hooke's Law relates the stresses to the strains:

$$\{\sigma\} = [D]\{\varepsilon\}. \tag{4.3}$$

For the two-dimensional continuum, the stress and strain vectors each contain three entries and the [D] matrix is a symmetrical 3×3 matrix.

Two dimensional problems are normally treated as plane stress or plane strain. The [D] matrices for these conditions are listed below.

For plane stress,

$$[D] = \frac{E}{1-v^2} \begin{bmatrix} 1 & v & 0 \\ v & 1 & 0 \\ 0 & 0 & \dfrac{1-v}{2} \end{bmatrix}$$

(4.4a)

For plane strain,

$$[D] = \frac{E}{(1+v)(1-2v)} \begin{bmatrix} 1-v & v & 0 \\ v & 1-v & 0 \\ 0 & 0 & \dfrac{1-2v}{2} \end{bmatrix}$$

(4.4b)

Often the plane stress formulation is used to model structures that are short in the third (out-of-plane) dimension, such as a circlip or a narrow, deep cantilever beam. The plane strain formulation is used to model structures that are long in the third dimension and that are constrained at the ends, such as the dam of a reservoir. In general, the analyst will determine which formulation is more appropriate by consideration of the potential of the boundary conditions to sustain load in the third dimension.

In addition to these two cases, structures that are axisymmetric can be treated using an extended form of the stress/strain relationship of the two-dimensional continuum. Although axisymmetric structures are three-dimensional they can be modelled very efficiently using two-dimensional elements in much the same way as for plane stress and plane strain cases. This is because the strain and the stress in the hoop direction can be related directly to the radial displacement. An extra term representing the hoop component is added to the strain and stress vectors but no additional nodal degrees of freedom are required. In this case the general three-dimensional expression of Hooke's Law reduces to:

$$\begin{Bmatrix} \sigma_x \\ \sigma_y \\ \sigma_z \\ \tau_{xy} \end{Bmatrix} = \frac{E}{(1+v)(1-2v)} \begin{bmatrix} 1-v & v & v & 0 \\ v & 1-v & v & 0 \\ v & v & 1-v & 0 \\ 0 & 0 & 0 & \dfrac{1-2v}{2} \end{bmatrix} \begin{Bmatrix} \varepsilon_x \\ \varepsilon_y \\ \varepsilon_z \\ \gamma_{xy} \end{Bmatrix}$$

(4.4c)

4.1.4 The principle of virtual displacements for the two-dimensional continuum

The principle of virtual displacements was stated in Section 3.1.2.

'When a small virtual displacement is imposed on a body that is in equilibrium, the additional strain energy generated in the body is equal to the virtual work done by the external forces acting on the body.'

The mathematical expression of this statement is re-iterated below.

$$\int_{Vol} \{\delta\ \varepsilon\}^{T} \{\sigma\}\ dVol = \int_{Vol} \{\delta\ u\}^{T} \{f\}_{body}\ dVol + \int_{Surf} \{\delta\ u\}^{T} \{f\}_{surf}\ dSurf + \{\delta\ u\}^{T}_{nodes} \{X\}_{nodes}$$

$$(4.5)$$

For the one-dimensional bar element, the right-hand side of the equation is reduced to two terms, associated with body forces and with nodal loads. In this case, there is an extra term that is attributable to the work done by surface pressures and tractions. Further, many of the quantities of the one-dimensional case are now expressed as matrices.

4.2 The Finite Element Formulation for a Continuum

The first step in the development of the finite element representation of the system of one-dimensional bar elements was to assume shape functions that related the displacement at the nodes of the elements to the displacements inside the elements. Appropriate shape functions were developed immediately (Section 3.2) so that the formulation of the finite element equations that followed could be illustrated more clearly. The two-dimensional problem will be approached in reverse order. Initially, for the purposes of this section, it is simply assumed that appropriate shape functions exist for a continuum element, and general finite element equations for the two-dimensional continuum are derived. Shape functions for specific element shapes are then developed in subsequent sections.

Assume that, within an element, the continuous displacement variable, $\{u\}$, can be expressed as the product of shape functions, N, and the nodal displacements, $\{u\}$.

$$\{u\} = [N] \{u\} \tag{4.6}$$

The shape function matrix, $[N]$, has 2 rows and as many columns as there are degrees of freedom in the element.

As for the one-dimensional system, the finite element formulation can be derived by appealing to the principle of virtual displacements, using the assumed displacements represented by the shape functions to approximate the strain energy within the elements. The strains can be written in terms of the displacements:

$$
\begin{Bmatrix} \varepsilon_x \\ \varepsilon_y \\ \gamma_{xy} \end{Bmatrix} =
\begin{bmatrix} \dfrac{\partial}{\partial x} & 0 \\ 0 & \dfrac{\partial}{\partial y} \\ \dfrac{\partial}{\partial y} & \dfrac{\partial}{\partial x} \end{bmatrix}
\begin{bmatrix} u \\ v \end{bmatrix} =
\begin{bmatrix} \dfrac{\partial}{\partial x} & 0 \\ 0 & \dfrac{\partial}{\partial y} \\ \dfrac{\partial}{\partial y} & \dfrac{\partial}{\partial x} \end{bmatrix} \{u\} =
\begin{bmatrix} \dfrac{\partial}{\partial x} & 0 \\ 0 & \dfrac{\partial}{\partial y} \\ \dfrac{\partial}{\partial y} & \dfrac{\partial}{\partial x} \end{bmatrix} [N]\{u\}
$$

or

$$\{\varepsilon\} = [B]\{u\} \tag{4.7}$$

where $\{\varepsilon\}$ is the continuous strain vector, evaluable at all points in the element, $\{u\}$ is the vector of discrete nodal displacements, and

$$
[B] =
\begin{bmatrix} \dfrac{\partial}{\partial x} & 0 \\ 0 & \dfrac{\partial}{\partial y} \\ \dfrac{\partial}{\partial y} & \dfrac{\partial}{\partial x} \end{bmatrix} [N] \tag{4.8}
$$

The matrix $[B]$ has three rows and as many columns as the element has degrees of freedom.

Now the global system equations can be written down by the application of small virtual displacements to the nodes when the structure is in equilibrium under all surface pressures, tractions, body forces and real nodal loads (it is assumed that there is a node at any point of application of an external force). By the principle of virtual displacements (Equation (4.5)),

$$\int_{Vol} \{\delta\,\varepsilon\}^T \{\sigma\}\ dVol = \int_{Vol} \{\delta\,u\}^T \{W\}\,dVol + \int_{Surf} \{\delta\,u\}^T \{p\}\,dSurf + \{\delta\,u\}^T \{X\} \quad (4.8.1)$$

where $\{W\}$ is the vector of body forces per unit volume, $\{p\}$ is the vector of pressures on the element surfaces and $\{X\}$ is the vector of nodal forces. The integrals over the structure are expressed as summations over the elements.

$$\sum_{elems} \int_{Vol} \{\delta\,\varepsilon\}^T \{\sigma\}\ dVol = \sum_{elems} \left(\int_{Vol} \{\delta\,u\}^T \{W\}\,dVol + \int_{Surf} \{\delta\,u\}^T \{p\}\,dSurf \right) + \{\delta\,u\}^T \{X\}$$

$$(4.8.2)$$

The displacement vector is related to the discrete nodal displacement vector by Equation (4.6), the strain and virtual strain vectors are related to the discrete nodal displacement and virtual displacement vectors by Equation (4.7), and the stress vector is related to the strain vector by Equation (4.3).

$$\sum_{elems} \int_{Vol} \{\delta\,u\}^T [B]^T [D][B]\{u\}\ dVol = \sum_{elems} \left(\int_{Vol} \{\delta\,u\}^T [N]^T \{W\}dVol + \int_{Surf} \{\delta\,u\}^T [N]^T \{p\}dSurf \right) + \{\delta\,u\}^T \{X\}$$

$$(4.8.3)$$

The virtual displacement vector is independent of the integrals and can be taken outside them.

$$\{\delta\,u\}^T \left(\sum_{elems} \int_{Vol} [B]^T [D][B]\ dVol \right) \{u\} = \{\delta\,u\}^T \left(\sum_{elems} \int_{Vol} [N]^T \{W\}dVol + \int_{Surf} [N]^T \{p\}dSurf \right) + \{\delta\,u\}^T \{X\}$$

Since this equation holds for all arbitrary values of the virtual displacement, the coefficients of each term in the virtual displacement vector must be the same.

$$[k]\{u\} = \{F\} \quad (4.9)$$

where

$$[k] = \sum_{elems} \int_{Vol} [B]^T [D][B]dVol = \sum_{elems} [k]_{elem}. \quad (4.10)$$

and

$$\{F\} = \sum_{elems} \left(\int_{Vol} [N]^T \{W\} dVol + \int_{Surf} [N]^T \{p\} dSurf \right) + \{X\} \tag{4.11}$$

The square matrix [k] is the global stiffness matrix and the vector $\{F\}$ is the global force vector, of which the nodal force vector $\{X\}$ is only one part.

The element stiffness matrices, $[k]_{elem}$, can be calculated by integration of a quantity featuring products of the derivatives of the shape functions and the Hookean stress/strain matrix over the volumes of the elements:

$$[k]_{elem} = \int_{Vol} [B]^T [D][B].dVol \tag{4.11.1}$$

4.3 A Triangular Element

4.3.1 Shape functions

Simple linear shape functions for a one-dimensional bar element were developed in Section 3.2. Similar functions can be developed for a triangular element. Assume now that there is a third node off the line of the first two, forming a triangular two-dimensional continuum element.

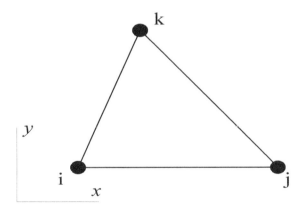

Figure 4.2 Triangular two-dimensional continuum element.

The force vector at node i is $\{X\}_i = \{X_i, Y_i\}^T$ and the displacement vector is $\{u\}_i = \{u_i, v_i\}^T$. Similar vectors apply at nodes j and k. The shape functions N_i, N_j and N_k will now be functions of two variables, i.e., x and y.

Now assume that the displacements anywhere in the element can be expressed as functions of the displacements of the nodes. The functions must satisfy the boundary conditions, $u = u_i$ and $v = v_i$ at node i and similar conditions at nodes j and k. This requirement will be satisfied if functions of the form:

$$u = N_i u_i + N_j u_j + N_k u_k \tag{4.12a}$$

$$v = N_i v_i + N_j v_j + N_k v_k \tag{4.12b}$$

are chosen, in which the shape function N_i adopts a value of zero at $(x = x_j, y = y_j)$ and at $(x = x_k, y = y_k)$, and of one at $(x = x_i, y = y_i)$. Similar functions are chosen for N_j and N_k.

In matrix form:

$$\begin{bmatrix} u \\ v \end{bmatrix} = \begin{bmatrix} N_i & 0 & N_j & 0 & N_k & 0 \\ 0 & N_i & 0 & N_j & 0 & N_k \end{bmatrix} \begin{bmatrix} u_i \\ v_i \\ u_j \\ v_j \\ u_k \\ v_k \end{bmatrix} \tag{4.13}$$

For the bar element, the most obvious choices of shape function were those that implied a linear variation of the displacement between the nodes. A similar function can be chosen for the triangular element, but in this case, there is an extra term representing the variation in the y direction. Consider the shape function N_k.

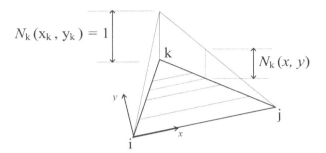

Figure 4.3 Triangular element shape function.

Now the magnitude of the shape function at any point is determined by the distance of the point from the line connecting nodes *i* and *j*. In a local co-ordinate system defined with the *x* axis lying along this line, this is easily written down.

$$N_k = \frac{y}{y_k}.$$

If the x axis is not co-incident with the line *ij*, but is parallel to it and displaced by a distance y_i, a more general formula is:

$$N_k = \frac{y - y_i}{y_k - y_i}.$$

Finally, if the triangular element is randomly orientated, with the line *ij* at an angle θ to the global *x* axis,

$$N_k = \frac{(y\cos\theta - x\sin\theta) - (y_i\cos\theta - x_i\sin\theta)}{(y_k\cos\theta - x_k\sin\theta) - (y_i\cos\theta - x_i\sin\theta)}.$$

Dividing every term on the rhs by cos θ and noting that tan $\theta = (y_j-y_i)/(x_j-x_i)$, and by expanding:

$$N_k = \frac{(x_i\,y_j - x_j\,y_i) + (y_i - y_j)x + (-x_i + x_j)y}{x_i(y_j - y_k) + x_j(-y_i + y_k) + x_k(y_i - y_j)}$$

The denominator in this expression will be recognised as twice the area of the triangle. Then:

$$N_k = \alpha_{k,1} + \alpha_{k,x} x + \alpha_{k,y} y \qquad (4.14)$$

where

$$\alpha_{k,1} = (x_i y_j - x_j y_i)/2A \quad : \quad \alpha_{k,x} = (y_i - y_j)/2A \quad : \quad \alpha_{k,y} = (-x_i + x_j)/2A$$

and A is the area of the triangle.

The other two shape functions, N_i and N_j, can be written in similar form.

$$N_i = \alpha_{i,1} + \alpha_{i,x} x + \alpha_{i,y} y$$

$$\alpha_{i,1} = (x_j y_k - x_k y_j)/2A \quad : \quad \alpha_{i,x} = (y_j - y_k)/2A \quad : \quad \alpha_{i,y} = (-x_j + x_k)/2A$$

$$N_j = \alpha_{j,1} + \alpha_{j,x} x + \alpha_{j,y} y$$

$$\alpha_{j,1} = (x_k y_i - x_i y_k)/2A \quad : \quad \alpha_{j,x} = (y_k - y_i)/2A \quad : \quad \alpha_{i,y} = (-x_k + x_i)/2A$$

Note that the α are simply constants that are derivable from the original geometry of the triangle, and that every triangle will therefore have its own shape functions depending on its position in the continuum.

4.3.2 The element stiffness matrix

For this triangular element, the expanded form of Equation (4.7) is:

$$
\begin{bmatrix} \varepsilon_x \\ \varepsilon_y \\ \gamma_{xy} \end{bmatrix}
=
\begin{bmatrix} \dfrac{\partial}{\partial x} & 0 \\ 0 & \dfrac{\partial}{\partial y} \\ \dfrac{\partial}{\partial y} & \dfrac{\partial}{\partial x} \end{bmatrix}
\begin{bmatrix} u \\ v \end{bmatrix}
=
\begin{bmatrix} \dfrac{\partial}{\partial x} & 0 \\ 0 & \dfrac{\partial}{\partial y} \\ \dfrac{\partial}{\partial y} & \dfrac{\partial}{\partial x} \end{bmatrix}
\begin{bmatrix} N_i & 0 & N_j & 0 & N_k & 0 \\ 0 & N_i & 0 & N_j & 0 & N_k \end{bmatrix}
\begin{bmatrix} u_i \\ v_i \\ u_j \\ v_j \\ u_k \\ v_k \end{bmatrix}
= [B][u]
$$

where

$$[B] = \begin{bmatrix} \dfrac{\partial N_i}{\partial x} & 0 & \dfrac{\partial N_j}{\partial x} & 0 & \dfrac{\partial N_k}{\partial x} & 0 \\[2ex] 0 & \dfrac{\partial N_i}{\partial y} & 0 & \dfrac{\partial N_j}{\partial y} & 0 & \dfrac{\partial N_k}{\partial y} \\[2ex] \dfrac{\partial N_i}{\partial y} & \dfrac{\partial N_i}{\partial x} & \dfrac{\partial N_j}{\partial y} & \dfrac{\partial N_j}{\partial x} & \dfrac{\partial N_k}{\partial y} & \dfrac{\partial N_k}{\partial x} \end{bmatrix}.$$

Noting the form of the shape functions (Equation (4.14)), and performing the differentiations,

$$[B] = \frac{1}{2A} \begin{bmatrix} y_j - y_k & 0 & y_k - y_i & 0 & y_i - y_j & 0 \\ 0 & x_k - x_j & 0 & x_i - x_k & 0 & x_j - x_i \\ x_k - x_j & y_j - y_k & x_i - x_k & y_k - y_i & x_j - x_i & y_i - y_j \end{bmatrix} \qquad (4.15)$$

It is important to note that, for this element, the entries in [B] are all constants, depending only upon the geometry and position of the triangle. This means that for any particular set of nodal displacements the three strain quantities are all constant over the triangle (see Equation (4.7)). This type of element is, therefore, often referred to as a **constant strain triangle.**

Now all of the elements of Equation (4.11.1) are known. Assuming that the element is of constant thickness (t), the integration $dVol$ can be replaced by t dA. Hence:

$$[k]_{elem} = \int_{Vol} [B]^T [D][B] \, t \, dA$$

But the matrices [B] and [D] contain only constants (i.e., they are not dependent on the co-ordinates x and y within the element). If t is constant, all of the terms can be taken outside the integral sign, and the integral is then simply the volume of the element itself.

$$[k]_{elem} = [B]^T [D][B] \, t \, A \qquad (4.16)$$

Equation (4.16) can be used in combination with Equations (4.15) and one of Equations (4.4a) or (4.4b), to derive the stiffness matrix of any three node triangular element of a continuum, assuming linear element shape functions.

4.3.3 Body forces

The global force vector (see Equation (4.11)) contains contributions from the body forces acting within the element. For structural analyses, the most common body force is that due to the action of gravity. The contribution to the global force vector from one element is:

$$\{X\}_{body} = \int_{Vol} [N]^T \{W\} \, dVol \,.$$

Assuming that the body force per unit volume, W, is constant in each of the two directions, and that the thickness of the triangle, t, is also constant:

$$\{X\}_{body} = \int_{Vol} [N]^T \{W\} t \, dA = \int_{Area} \begin{bmatrix} N_i & 0 \\ 0 & N_i \\ N_j & 0 \\ 0 & N_j \\ N_k & 0 \\ 0 & N_k \end{bmatrix} dA \begin{bmatrix} W_x \\ W_y \end{bmatrix} t$$

The integral over the area of the element of each of the shape functions is required. This can be written down by consideration of the following figure.

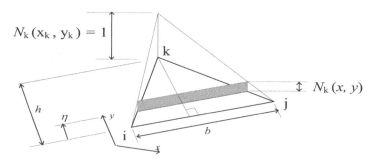

$N_k (x_k, y_k) = 1$

$N_k (x, y)$

Figure 4.4 Integration of triangular element shape function.

The volume shaded is

$$dV = \left(\frac{b(h-\eta)}{h}\right) \cdot \left(\frac{\eta}{h}\right) \cdot d\eta ,$$

and the total volume is

$$V = \int_0^h \left(\frac{b(h-\eta)}{h}\right) \cdot \left(\frac{\eta}{h}\right) \cdot d\eta = \frac{bh}{6} = \frac{A}{3} ,$$

where A is the area of the triangle.

$$\{X\}_{body} = \int_{Area} \begin{bmatrix} N_i & 0 \\ 0 & N_i \\ N_j & 0 \\ 0 & N_j \\ N_k & 0 \\ 0 & N_k \end{bmatrix} dA \begin{bmatrix} W_x \\ W_y \end{bmatrix} t = \frac{At}{3} \begin{bmatrix} W_x \\ W_y \\ W_x \\ W_y \\ W_x \\ W_y \end{bmatrix} \tag{4.17}$$

Since the total body force in the x direction is (AtW_x), this implies that one third of the body force should be applied at each of the nodes of the element.

4.3.4 Surface pressures and tractions

The global force vector also contains contributions from pressures or tractions acting on the surfaces of the element.

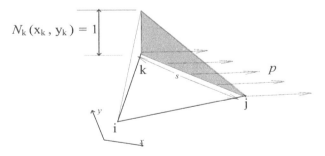

Figure 4.5

Assume that the element has a line load per unit length with components w_x and w_y applied to the side *jk*, and that the length of side *jk* is s. The pressure on the surface has components p_x and p_y, calculated by dividing the line load by the element thickness.

$$\{X\}_{surf} = \int_{Surf} [N]^T \{p\} \, dSurf = \int_{jk} \begin{bmatrix} N_i & 0 \\ 0 & N_i \\ N_j & 0 \\ 0 & N_j \\ N_k & 0 \\ 0 & N_k \end{bmatrix} t \, ds \begin{bmatrix} p_x \\ p_y \end{bmatrix}$$

Now the integral over the side *jk* of each of the shape functions is required.

$$\int_{jk} N_i \, dSurf = 0 \quad : \quad \int_{jk} N_j \, dSurf = \frac{ts}{2} \quad : \quad \int_{jk} N_k \, dSurf = \frac{ts}{2}$$

The element force vector consistent with the pressure on side *jk* is, therefore:

$$\{X\}_{surf} = \int_{jk} \begin{bmatrix} N_i & 0 \\ 0 & N_i \\ N_j & 0 \\ 0 & N_j \\ N_k & 0 \\ 0 & N_k \end{bmatrix} t \, ds \begin{bmatrix} p_x \\ p_y \end{bmatrix} = \frac{ts}{2} \begin{bmatrix} 0 \\ 0 \\ p_x \\ p_y \\ p_x \\ p_y \end{bmatrix} \qquad (4.18)$$

where s is the length of side *jk*.

Hence, the appropriate treatment of a pressure force on the side of the element would be to distribute it equally between the nodes on that side. For both the body force treated in the previous section and the edge force treated in this section, the analysis showed that the obvious distribution of the forces to the nodes is consistent with the element formulation. It will be demonstrated later that this is not the case for the higher order elements, and care should be taken when specifying distributed loads in real finite element analyses.

4.3.5 *Numerical example: A single element*

Write down the [B] and [D] matrices and calculate the stiffness matrix and the load vector for the plane-stress system illustrated below when it is modelled by a single constant strain triangular element. Solve for the displacements of the structure and calculate the strains and then the stresses. The thickness of the membrane is 2 mm, and it is made of aluminium with a Young's modulus of 70 kN/mm² and a Poisson's ratio of 0.30.

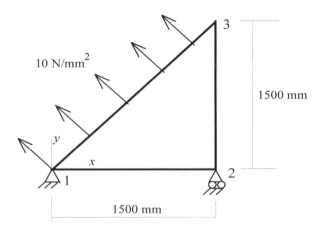

The first step in the solution is to write down the co-ordinates of the nodes:

Node	x (mm)	y (mm)
1	0	0
2	1500	0
3	1500	1500

The area of the triangle, A, is

$$A = \frac{1}{2} \cdot 1500 \cdot 1500 = 1.125 \text{ x } 10^6 \text{ mm}^2.$$

The [B] matrix can now be written down from Equation (5.15).

$$[B] = \frac{1}{2.25 \times 10^6} \begin{bmatrix} -1500 & 0 & 1500 & 0 & 0 & 0 \\ 0 & 0 & 0 & -1500 & 0 & 1500 \\ 0 & -1500 & -1500 & 1500 & 1500 & 0 \end{bmatrix}$$

The [D] matrix for plane stress is:

$$[D] = \frac{E}{\left(1-v^2\right)} \begin{bmatrix} 1 & v & 0 \\ v & 1 & 0 \\ 0 & 0 & \frac{1-v}{2} \end{bmatrix} = 76923 \begin{bmatrix} 1 & 0.3 & 0 \\ 0.3 & 1 & 0 \\ 0 & 0 & 0.35 \end{bmatrix}$$

The matrix [k] is formed by matrix multiplication (Equation 4.16)

$$[B]^T[D] = \frac{76923}{2.25 \times 10^6} \begin{bmatrix} -1500 & 0 & 0 \\ 0 & 0 & -1500 \\ 1500 & 0 & -1500 \\ 0 & -1500 & 1500 \\ 0 & 0 & 1500 \\ 0 & 1500 & 0 \end{bmatrix} \begin{bmatrix} 1 & 0.3 & 0 \\ 0.3 & 1 & 0 \\ 0 & 0 & 0.35 \end{bmatrix} = 0.0342 \begin{bmatrix} -1500 & -450 & 0 \\ 0 & 0 & -525 \\ 1500 & 450 & -525 \\ -450 & -1500 & 525 \\ 0 & 0 & 525 \\ 450 & 1500 & 0 \end{bmatrix}$$

$$= \frac{0.0342 \times 10^6}{2.25 \times 10^6} \begin{bmatrix} 2.25 & 0 & -2.25 & 0.675 & 0 & -0.675 \\ 0 & 0.7875 & 0.7875 & -0.7875 & -0.7875 & 0 \\ -2.25 & 0.7875 & 3.0375 & -1.4635 & -0.7875 & 0.675 \\ 0.675 & -0.7875 & -1.4625 & 3.0375 & 0.7875 & -2.25 \\ 0 & -0.7875 & -0.7875 & 0.7875 & 0.7875 & 0 \\ -0.675 & 00 & 0.675 & -2.25 & 0 & 2.25 \end{bmatrix}$$

$$[k] = [B]^T [D][B] \, t \, A = [B]^T [D][B] \times 2 \times 1.125 \times 10^6$$

$$= \begin{bmatrix}
76950 & 0 & -76950 & 23085 & 0 & -23085 \\
0 & 26932.5 & 26932.5 & -26932.5 & -26932.5 & 0 \\
-76950 & 26932.5 & 103882.5 & -50017.5 & -26932.5 & 23085 \\
23085 & -26932.5 & -50017.5 & 103882.5 & 26932.5 & -76950 \\
0 & -26932.5 & -26932.5 & 26932.5 & 26932.5 & 0 \\
-23085 & 0 & 23085 & -76950 & 0 & 76950
\end{bmatrix}$$

The only load on the system is a pressure load, and the load vector can be calculated from Equation (4.18). Resolving the pressure into the x and y components:

$$p_x = 10 \text{ x} - \sin45 = -7.071 \ N/mm^2 \qquad p_y = 10 \text{ x} \cos45 = 7.071 \ N/mm^2$$

$$\{X\}_{surf} = \frac{ts}{2} \begin{bmatrix} p_x \\ p_y \\ 0 \\ 0 \\ p_x \\ p_y \end{bmatrix} = \frac{2 \times 2121}{2} \begin{bmatrix} -7.071 \\ 7.071 \\ 0 \\ 0 \\ -7.071 \\ 7.071 \end{bmatrix} = \begin{bmatrix} -15000 \\ 15000 \\ 0 \\ 0 \\ -15000 \\ 15000 \end{bmatrix}$$

Adding on the unknown reactions to the force vector, and inserting the boundary conditions into the displacement vector, the finite element equations ($\{F\} = [k]\{u\}$) for the system become:

$$\begin{bmatrix} -15000+X_1 \\ 15000+Y_1 \\ 0 \\ Y_2 \\ -15000 \\ 15000 \end{bmatrix} = \begin{bmatrix}
76950 & 0 & -76950 & 23085 & 0 & -23085 \\
0 & 26932.5 & 26932.5 & -26932.5 & -26932.5 & 0 \\
-76950 & 26932.5 & 103882.5 & -50017.5 & -26932.5 & 23085 \\
23085 & -26932.5 & -50017.5 & 103882.5 & 26932.5 & -76950 \\
0 & -26932.5 & -26932.5 & 26932.5 & 26932.5 & 0 \\
-23085 & 0 & 23085 & -76950 & 0 & 76950
\end{bmatrix} \begin{bmatrix} 0 \\ 0 \\ u_2 \\ 0 \\ u_3 \\ v_3 \end{bmatrix}.$$

Eliminating the rows and columns associated with the zero boundary conditions:

$$
\begin{bmatrix} 0 \\ -15000 \\ 15000 \end{bmatrix} = \begin{bmatrix} 103882.5 & -26932.5 & 23085 \\ -26932.5 & 26932.5 & 0 \\ 23085 & 0 & 76950 \end{bmatrix} \begin{bmatrix} u_2 \\ u_3 \\ v_3 \end{bmatrix}
$$

The solution now proceeds by using Gaussian elimination to reduce the matrix to upper triangular. This procedure is shown in detail in Section 2.2.4. Subtracting (23085/103882.5) times the first row from the third row will eliminate the entry in the third row of the first column. Similarly subtracting (−26932.5/103882.5) times the first row from the second row will eliminate the entry in the second row of the first column.

$$
\begin{bmatrix} 0 \\ -15000 \\ 15000 \end{bmatrix} = \begin{bmatrix} 103882.5 & -26932.5 & 23085 \\ 0 & 19950 & 5985 \\ 0 & 5985 & 71820 \end{bmatrix} \begin{bmatrix} u_2 \\ u_3 \\ v_3 \end{bmatrix}
$$

Eliminating the entry in the third row of the second column,

$$
\begin{bmatrix} 0 \\ -15000 \\ 19500 \end{bmatrix} = \begin{bmatrix} 103882.5 & -26932.5 & 23085 \\ 0 & 19950 & 5985 \\ 0 & 0 & 70024.5 \end{bmatrix} \begin{bmatrix} u_2 \\ u_3 \\ v_3 \end{bmatrix}
$$

The solution for the displacements can be found by back-substitution.

$$
v_3 = \frac{19500}{70024.5} = 0.278 \ \text{mm}
$$

$$
u_3 = \frac{1}{19950}(-15000 - 5985v_3) = -0.835 \ \text{mm}
$$

$$
u_2 = \frac{1}{103882.5}(0 + 26932.5u_3 - 23085v_3) = -0.278 \ \text{mm}
$$

The reactions can be recovered from the global equation.

$$X_1 = -76950u_2 - 0u_3 - 23085v_3 + 15000 = 30000 \text{ N}$$

$$Y_1 = 26932.5u_2 - 26932.5u_3 - 0v_3 - 15000 = 0 \text{ N}$$

$$Y_2 = -50017.5u_2 + 26932.5u_3 - 76950v_3 = -30000 \text{ N}$$

The strains can be found from Equation (4.7).

$$\begin{bmatrix} \varepsilon_x \\ \varepsilon_y \\ \gamma_{xy} \end{bmatrix} = [B]\{u\} = \frac{1}{2.25 \times 10^6} \begin{bmatrix} -1500 & 0 & 1500 & 0 & 0 & 0 \\ 0 & 0 & 0 & -1500 & 0 & 1500 \\ 0 & -1500 & -1500 & 1500 & 1500 & 0 \end{bmatrix} \begin{bmatrix} 0 \\ 0 \\ -0.278 \\ 0 \\ -0.835 \\ 0.278 \end{bmatrix} = 10^{-6} \begin{bmatrix} -186 \\ 186 \\ -371 \end{bmatrix}$$

The stresses can be found from Equations (4.3) and (4.4a):

$$\begin{bmatrix} \sigma_x \\ \sigma_y \\ \tau_{xy} \end{bmatrix} = [D] \begin{bmatrix} \varepsilon_x \\ \varepsilon_y \\ \gamma_{xy} \end{bmatrix} = 76923 \times 1 \times 10^{-6} \begin{bmatrix} 1 & 0.3 & 0 \\ 0.3 & 1 & 0 \\ 0 & 0 & 0.35 \end{bmatrix} \begin{bmatrix} -186 \\ 186 \\ -371 \end{bmatrix} = \begin{bmatrix} -10 \\ 10 \\ -10 \end{bmatrix}$$

As discussed in the development of the stiffness matrix, the strain (and therefore the stress) is constant over the whole of the element as a consequence of the form of the shape function. It is obvious that this would not be true in a real triangular membrane loaded and supported in this fashion, and the finite element analysis with a single element provides only a gross approximation of the stress distribution in this system.

4.4 A Quadrilateral Element

4.4.1 Shape functions

Simple linear shape functions for a triangular element were developed in Section 4.3.1. Similar functions can be developed for a quadrilateral element, but unless the axes of the quadrilateral are parallel to the global axes, the algebra becomes very clumsy. In fact, the best methods for development of quadrilateral elements are based on a definition of the element in an idealised square space with a mapping function relating points in this space to points in real space. This approach is referred to as a parametric element formulation. It is, however, instructive to consider the development of a quadrilateral element with axes parallel to the global axes using the methods applied to the triangular element.

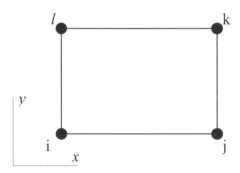

Figure 4.6 Quadrilateral element.

The force vector at node i is $\{X\}_i = \{X_i, Y_i\}^T$ and the displacement vector is $\{u\}_i = \{u_i, v_i\}^T$. Similar vectors apply at nodes j, k and l.

Now assume that the displacements anywhere in the element can be expressed as functions of the displacements of the nodes. The functions must satisfy the boundary conditions, $u = u_i$ and $v = v_i$ at node i and similar conditions at nodes j, k and l. This requirement will be satisfied if functions of the form

$$u = N_i u_i + N_j u_j + N_k u_k + N_l u_l \qquad (4.19a)$$

$$v = N_i v_i + N_j v_j + N_k v_k + N_l v_l \qquad (4.19b)$$

are chosen, in which the shape function N_i adopts a value of zero at $(x = x_j, y = y_j)$, at $(x = x_k, y = y_k)$, and at $(x = x_l, y = y_l)$, and of one at $(x = x_i, y = y_i)$. Similar functions are chosen for the other shape functions. Again, the obvious functions, implying a linear variation of the displacement between the nodes, are chosen.

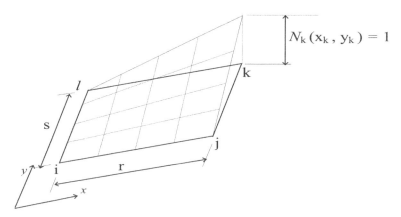

Figure 4.7 Quadrilateral element shape function.

$$N_k = \left(\frac{x - x_i}{r}\right)\left(\frac{y - y_i}{s}\right) = \frac{1}{A}\left((x - x_i)(y - y_i)\right)$$

$$N_k = \alpha_{k,1} + \alpha_{k,x}\,x + \alpha_{k,y}\,y + \alpha_{k,xy}\,x\,y. \tag{4.20}$$

where

$$\alpha_{k,1} = x_i\,y_i/A \quad : \quad \alpha_{k,x} = -y_i/A \quad : \quad \alpha_{k,y} = -x_i/A \quad : \quad \alpha_{k,xy} = 1/A$$

and A is the area of the quadrilateral.

The other three shape functions can be written in similar form. As for the constant strain triangle, the α are simply constants that are derivable from the original geometry of the quadrilateral, and every quadrilateral will therefore have its own shape functions depending on its position in the continuum.

4.4.2 The element stiffness matrix

For this quadrilateral element, the expanded form of Equation (4.7) is:

$$\begin{bmatrix} \varepsilon_x \\ \varepsilon_y \\ \gamma_{xy} \end{bmatrix} = \begin{bmatrix} \dfrac{\partial}{\partial x} & 0 \\ 0 & \dfrac{\partial}{\partial y} \\ \dfrac{\partial}{\partial y} & \dfrac{\partial}{\partial x} \end{bmatrix} \begin{bmatrix} u \\ v \end{bmatrix} = \begin{bmatrix} \dfrac{\partial}{\partial x} & 0 \\ 0 & \dfrac{\partial}{\partial y} \\ \dfrac{\partial}{\partial y} & \dfrac{\partial}{\partial x} \end{bmatrix} \begin{bmatrix} N_i & 0 & N_j & 0 & N_k & 0 & N_l & 0 \\ 0 & N_i & 0 & N_j & 0 & N_k & 0 & N_l \end{bmatrix} \begin{bmatrix} u_i \\ v_i \\ u_j \\ v_j \\ u_k \\ v_k \\ u_l \\ v_l \end{bmatrix} = [B]\{u\}$$

Noting the form of the shape functions (Equation (4.20)), and performing the differentiations,

$$[B] = \frac{1}{A}\begin{bmatrix} y-y_k & 0 & -(y-y_k) & 0 & y-y_i & 0 & -(y-y_i) & 0 \\ 0 & x-x_k & 0 & -(x-x_i) & 0 & x-x_i & 0 & -(x-x_k) \\ x-x_k & y-y_k & -(x-x_i) & -(y-y_k) & x-x_i & y-y_i & -(x-x_k) & -(y-y_i) \end{bmatrix}$$

(4.21)

Now, unlike for the triangle, the entries in [B] are dependent on the position within the element. The first row of [B] multiplied by the nodal displacement vector, {u}, gives the strain in the x direction. It can be seen that this component of strain, ε_x, varies linearly in the y direction but is constant in the x direction. This is because the first row of [B] has linear terms in y but no terms in x. Similarly, the strain in the y direction, ε_y, is constant in the y direction but varies linearly in the x direction. This element is sometimes called the constant stress quadrilateral, although the element formulation implies that the stress is not in fact constant over the area.

Assuming that the element is of constant thickness (t), integration with respect to volume, *dVol*, can be replaced by t dA. The element stiffness matrix for plane stress can be determined by substitution of Equations (4.4a) and (4.21) into Equation (4.11.1).

$$[k]_{elem} = \int_{Vol} [B]^T [D][B]\, t\, dA$$

Now the [D] matrix contains only constants but the [B] matrix contains terms that are dependent on the x and y co-ordinates, and the evaluation of the above integral is somewhat more difficult than it was for the triangular element.

The matrix $[B]^T[D]$ has eight rows and three columns. For plane stress (using Equation (4.4a)).

$$[B]^T[D] = \frac{E}{A(1-v^2)}\begin{bmatrix} (y-y_k) & v(y-y_k) & \frac{(1-v)}{2}(x-x_k) \\ v(x-x_k) & (x-x_k) & \frac{(1-v)}{2}(y-y_k) \\ \cdot & \cdot & \cdot \\ \cdot & \cdot & \cdot \\ \cdot & \cdot & \cdot \\ \cdot & \cdot & \cdot \end{bmatrix}.$$

$$[B]^T[D][B] = \frac{E}{A^2(1-v^2)}\begin{bmatrix} (y-y_k)^2 + \frac{(1-v)}{2}(x-x_k)^2 & \cdot & \cdot & \cdot & \cdot & \cdot & \cdot & \cdot \\ \frac{(1+v)}{2}(x-x_k)(y-y_k) & & \cdot & \cdot & \cdot & \cdot & \cdot & \cdot \\ \cdot & & & \cdot & \cdot & \cdot & \cdot & \cdot \\ \cdot & & & \cdot & \cdot & \cdot & \cdot & \cdot \\ \cdot & & & \cdot & \cdot & \cdot & \cdot & \cdot \\ \cdot & & & \cdot & \cdot & \cdot & \cdot & \cdot \end{bmatrix}$$

Hence, the first term in the stiffness matrix for the element is:

$$k_{11} = \iint_{Area} \frac{E}{A^2(1-v^2)}\left((y-y_k)^2 + \frac{(1-v)}{2}(x-x_k)^2\right).t\ .dx.dy$$

For each term in the stiffness matrix it will be necessary to integrate a linear or a quadratic polynomial with respect to x and to y. In practice, these terms will be evaluated using a numerical integration technique. Numerical integration to obtain the stiffness matrix is discussed later.

Consider the rectangular element illustrated below.

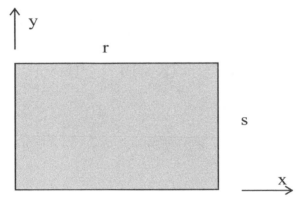

Figure 4.8 Rectangular element.

The first term in the stiffness matrix is:

$$k_{11} = \int_0^s \int_0^r \frac{E}{A^2(1-v^2)}\left[(y-y_k)^2 + \frac{(1-v)}{2}(x-x_k)^2\right] t\ dxdy = \frac{Et}{6rs\ (1-v^2)}\left(2s^2 + (1-v)r^2\right)$$

4.4.3 An application for the element

Consider the case of a long, deep beam subjected to a bending moment.

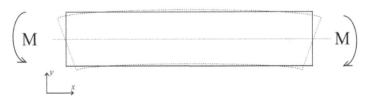

Figure 4.9 Beam subjected to bending moment.

According to bending theory and neglecting Poisson's ratio effects, the stress in the x direction, σ_x, will vary linearly in the y direction, but will be constant in the x direction. From this point of view the element developed in this section is ideal for this problem. It is anticipated that a model featuring a single element over the full depth of the beam would be accurate. Along the axis of the beam, the theory tells us that the curvature will be constant (for constant bending moment), and therefore the displacement in the y direction will be represented by a quadratic equation. This variation of displacement cannot be modelled by this element, and so it will be necessary to have sufficiently many elements along the axis of the beam to represent the quadratic as a series of straight lines. An appropriate mesh density for a model of this system might, therefore, be the one illustrated in Figure 4.10.

Figure 4.10 Mesh density illustrated on beam element.

The number of elements along the axis of the beam will be determined by the accuracy required from the analysis. Since the introduction of transverse shear forces in addition to bending moments simply changes the function representing y displacement to a cubic, which can also be represented in a piecewise linear fashion as above, this model will be appropriate for the study of the beam under more general loading conditions.

In practice, the Poisson's ratio effects introduce a linear variation of the y-direction strain across the depth of the beam, and this cannot be modelled by this element. In a real analysis, using this element the analyst would use a model with a few elements to cater for the variation caused by the Poisson's ratio terms.

4.5 Numerical Study: Pin-jointed Frame with a Shear Web

In Section 2.5.4, a space-frame internal wing structure of a light kit-form aircraft was analysed. Consider now a similar structure, but with the diagonal members in the rectangular portion replaced by a shear web.

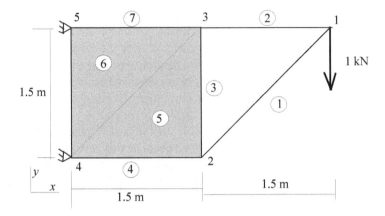

The framework is connected to a very stiff piece of structure at nodes 4 and 5, and the displacements of each of these nodes can be considered to be zero in both *x* and *y* directions. Each of the rod members in the framework has an extensional stiffness $AE/L = 10$ kN/mm. The shear web has a thickness of 2 mm and is made of aluminium with a Young's modulus of 70 kN/mm² and a Poisson's ratio of 0.3. A load of 1 kN is applied in the negative y direction at joint 1 as illustrated.

Calculate the vertical displacement of the loaded joint and the distribution of shear stress in the web. Calculate also the reactions in the vertical direction at the supports.

This analysis of this structure illustrates two important points.

- It features elements of different types, in this case bars and shear webs.
- The distribution of stress in the shear web will be complex, and the finite element analysis will not yield exact results as it did for the framework alone.

There is no problem in mixing element types in an analysis model, provided that the nodal degrees of freedom are compatible. Both the bar elements developed in Chapter 3 and the membrane elements developed in this section are defined in terms of the nodal translations, u and v. They can therefore be used together without difficulty. The use of elements with incompatible freedoms, such as beams and membranes, will be discussed later.

The shear web can be modelled using the membrane elements developed in this section. For the purposes of this analysis, it will be assumed that the web is connected to the bars only at the four corners. One possible model of the rectangular shear web is as two constant strain triangular elements as illustrated in the diagram. This analysis is worked through below in order to illustrate how the terms from the different elements combine in the global stiffness matrix, and to show how the strains and stresses can be calculated in a post-processing operation.

The first step in the solution procedure is to write down the element stiffness matrices. The stiffness matrices of the bars have already been calculated in example 2.5.4, and they are repeated below.

Element 1 AE/L = 10000 N/mm: $\theta = -135°$: $\sin\theta = -0.7071$ $\cos\theta = -0.7071$

$$
\begin{bmatrix} X_1 \\ Y_1 \\ X_2 \\ Y_2 \end{bmatrix}_1 = \begin{bmatrix} 5000 & 5000 & -5000 & -5000 \\ 5000 & 5000 & -5000 & -5000 \\ -5000 & -5000 & 5000 & 5000 \\ -5000 & -5000 & 5000 & 5000 \end{bmatrix} \begin{bmatrix} u_1 \\ v_1 \\ u_2 \\ v_2 \end{bmatrix}
$$

Element 2 AE/L = 10000 N/mm: $\theta = 180°$: $\sin\theta = 0$ $\cos\theta = -1$

$$
\begin{bmatrix} X_1 \\ Y_1 \\ X_3 \\ Y_3 \end{bmatrix}_2 = \begin{bmatrix} 10000 & 0 & -10000 & 0 \\ 0 & 0 & 0 & 0 \\ -10000 & 0 & 10000 & 0 \\ 0 & 0 & 0 & 0 \end{bmatrix} \begin{bmatrix} u_1 \\ v_1 \\ u_3 \\ v_3 \end{bmatrix}
$$

Element 3 AE/L = 10000 N/mm: $\theta = 90°$: $\sin\theta = 1$ $\cos\theta = -0.7071$

$$
\begin{bmatrix} X_2 \\ Y_2 \\ X_3 \\ Y_3 \end{bmatrix}_3 = \begin{bmatrix} 0 & 0 & 0 & 0 \\ 0 & 10000 & 0 & -10000 \\ 0 & 0 & 0 & 0 \\ 0 & -10000 & 0 & 10000 \end{bmatrix} \begin{bmatrix} u_2 \\ v_2 \\ u_3 \\ v_3 \end{bmatrix}
$$

Element 4 AE/L = 10000 N/mm: $\theta = 180°$: $\sin\theta = 0$ $\cos\theta = -1$

$$
\begin{bmatrix} X_2 \\ Y_2 \\ X_4 \\ Y_4 \end{bmatrix}_1 = \begin{bmatrix} 10000 & 0 & -10000 & 0 \\ 0 & 0 & 0 & 0 \\ -10000 & 0 & 10000 & 0 \\ 0 & 0 & 0 & 0 \end{bmatrix} \begin{bmatrix} u_2 \\ v_2 \\ u_4 \\ v_4 \end{bmatrix}
$$

Element 7 AE/L = 10000 N/mm: $\theta = 180°$: $\sin\theta = 0$ $\cos\theta = -1$

$$
\begin{bmatrix} X_3 \\ Y_3 \\ X_5 \\ Y_5 \end{bmatrix}_7 = \begin{bmatrix} 10000 & 0 & -10000 & 0 \\ 0 & 0 & 0 & 0 \\ -10000 & 0 & 10000 & 0 \\ 0 & 0 & 0 & 0 \end{bmatrix} \begin{bmatrix} u_3 \\ v_3 \\ u_5 \\ v_5 \end{bmatrix}
$$

The stiffness matrix for triangular element 5 was calculated in the previous example.

$$
\begin{bmatrix} X_4 \\ Y_4 \\ X_2 \\ Y_2 \\ X_3 \\ Y_3 \end{bmatrix} = \begin{bmatrix} 76950 & 0 & -76950 & 23085 & 0 & -23085 \\ 0 & 26932.5 & 26932.5 & -26932.5 & -26932.5 & 0 \\ -76950 & 26932.5 & 103882.5 & -50017.5 & -26932.5 & 23085 \\ 23085 & -26932.5 & -50017.5 & 103882.5 & 26932.5 & -76950 \\ 0 & -26932.5 & -26932.5 & 26932.5 & 26932.5 & 0 \\ -23085 & 0 & 23085 & -76950 & 0 & 76950 \end{bmatrix} \begin{bmatrix} u_4 \\ v_4 \\ u_2 \\ v_2 \\ u_3 \\ v_3 \end{bmatrix}
$$

The stiffness matrix for triangular element 6 can be calculated in a similar manner.

$$
\begin{bmatrix} X_3 \\ Y_3 \\ X_5 \\ Y_5 \\ X_4 \\ Y_4 \end{bmatrix} = \begin{bmatrix} 76950 & 0 & -76950 & 23085 & 0 & -23085 \\ 0 & 26932.5 & 26932.5 & -26932.5 & -26932.5 & 0 \\ -76950 & 26932.5 & 103882.5 & -50017.5 & -26932.5 & 23085 \\ 23085 & -26932.5 & -50017.5 & 103882.5 & 26932.5 & -76950 \\ 0 & -26932.5 & -26932.5 & 26932.5 & 26932.5 & 0 \\ -23085 & 0 & 23085 & -76950 & 0 & 76950 \end{bmatrix} \begin{bmatrix} u_3 \\ v_3 \\ u_5 \\ v_5 \\ u_4 \\ v_4 \end{bmatrix}
$$

The element stiffness matrices are now assembled into a global stiffness matrix for the structure.

$$
\begin{bmatrix}
5000{+}10000 & 5000{+}0 & -5000 & -5000 & -10000 & 0 & 0 & 0 & 0 & 0 \\
5000{+}0 & 5000{+}0 & -5000 & -5000 & 0 & 0 & 0 & 0 & 0 & 0 \\
-5000 & -5000 & 5000{+}0{+}10000{+}103883 & 5000{+}0{+}0{-}50018 & -26933 & 23085 & -10000{-}76950 & 26933 & 0 & 0 \\
-5000 & -5000 & 5000{+}0{+}0{-}50018 & 5000{+}10000{+}0{+}103883 & 26933 & -10000{-}76950 & 23085 & -26933 & 0 & 0 \\
-10000 & 0 & -26933 & 26933 & 10000{+}0{+}26933{+}76950{+}10000 & 0 & 0 & 0 & 23085 & -26933 \\
0 & 0 & 23085 & -10000{-}76950 & 0 & 0{+}10000{+}76950{+}26933{+}0 & -23085{-}26933 & 0 & 26933 & -26933 \\
0 & 0 & -10000{-}76950 & 23085 & 0 & -23085{-}26933 & 0{+}0{+}26933{+}76950{+}10000 & 0 & -26933 & 26933 \\
0 & 0 & 26933 & -26933 & 0 & 0 & 0 & 103883{+}0 & 23085 & -76950 \\
0 & 0 & 0 & 0 & 23085 & 26933 & -26933 & 23085 & 103883{+}10000 & -23085{-}26933 \\
0 & 0 & 0 & 0 & -26933 & -26933 & 26933 & -76950 & -23085{-}26933 & 103883{+}0
\end{bmatrix}
$$

Note that there are as many contributors to each of the entries on the main diagonal as there are members joining at the node under consideration. Adding the contributions together, the global stiffness matrix is calculated. At this stage, the known values of force and displacement at the nodes are entered into the force and displacement vectors.

$$
\begin{bmatrix}
0 \\ -1000 \\ 0 \\ 0 \\ 0 \\ 0 \\ X_4 \\ Y_4 \\ X_5 \\ Y_5
\end{bmatrix}
=
\begin{bmatrix}
15000 & 5000 & -5000 & -5000 & -10000 & 0 & 0 & 0 & 0 & 0 \\
5000 & 5000 & -5000 & -5000 & 0 & 0 & 0 & 0 & 0 & 0 \\
-5000 & -5000 & 118883 & -45018 & -26933 & 23085 & -86950 & 26933 & 0 & 0 \\
-5000 & -5000 & -45018 & 118883 & 26933 & -86950 & 23085 & -26933 & 0 & 0 \\
-10000 & 0 & -26933 & 26933 & 123883 & 0 & 0 & 0 & 23085 & -26933 \\
0 & 0 & 23085 & -86950 & 0 & 113883 & -50018 & 0 & 26933 & -26933 \\
0 & 0 & -86950 & 23085 & 0 & -50018 & 113883 & 0 & -26933 & 26933 \\
0 & 0 & 26933 & -26933 & 0 & 0 & 0 & 103883 & 23085 & -76950 \\
0 & 0 & 0 & 0 & 23085 & 26933 & -26933 & 23085 & 113883 & -50018 \\
0 & 0 & 0 & 0 & -26933 & -26933 & 26933 & -76950 & -50018 & 103883
\end{bmatrix}
\begin{bmatrix}
u_1 \\ v_1 \\ u_2 \\ v_2 \\ u_3 \\ v_3 \\ 0 \\ 0 \\ 0 \\ 0
\end{bmatrix}
$$

The first six equations contain the six unknown components of displacement (degrees of freedom) and all known forces. The equations are all independent, so they can be solved for displacement.

$$\begin{bmatrix} 0 \\ -1000 \\ 0 \\ 0 \\ 0 \\ 0 \end{bmatrix} = \begin{bmatrix} 15000 & 5000 & -5000 & -5000 & -10000 & 0 \\ 5000 & 5000 & -5000 & -5000 & 0 & 0 \\ -5000 & -5000 & 118883 & -45018 & -26933 & 23085 \\ -5000 & -5000 & -45018 & 118883 & 26933 & -86950 \\ -10000 & 0 & -26933 & 26933 & 123883 & 0 \\ 0 & 0 & 23085 & -86950 & 0 & 113883 \end{bmatrix} \begin{bmatrix} u_1 \\ v_1 \\ u_2 \\ v_2 \\ u_3 \\ v_3 \end{bmatrix}$$

Once again, the standard Gaussian elimination method can be used to reduce the matrix to upper triangular. The end result is:

$$\begin{bmatrix} 0 \\ -1000 \\ -1000 \\ -1439 \\ 1000 \\ -1172 \end{bmatrix} = \begin{bmatrix} 15000 & 5000 & -5000 & -5000 & -10000 & 0 \\ 0 & 3333 & -3333 & -3333 & 3333 & 0 \\ 0 & 0 & 113883 & -50018 & -26933 & 23085 \\ 0 & 0 & 0 & 91915 & 15103 & -76811 \\ 0 & 0 & 0 & 0 & 105031 & 18081 \\ 0 & 0 & 0 & 0 & 0 & 41902 \end{bmatrix} \begin{bmatrix} u_1 \\ v_1 \\ u_2 \\ v_2 \\ u_3 \\ v_3 \end{bmatrix}$$

The solution for the displacements can be determined by back-substitution.

$$v_3 = \frac{-1172}{41902} = -0.0280 \text{ mm}$$

$$u_3 = \frac{1}{105031}(1000 - 18081v_3) = 0.0143 \text{ mm}$$

$$v_2 = \frac{1}{91915}(-1439 - 15103u_3 + 76811v_3) = -0.0414 \text{ mm}$$

$$u_2 = \frac{1}{113883}(-1000 + 50018v_2 + 26933u_3 - 23085v_3) = -0.0180 \text{ mm}$$

$$v_1 = \frac{1}{3333}\left(-1000 + 3333u_2 + 3333v_2 - 3333u_3 - 0v_3\right) = -0.3736 \text{ mm}$$

$$u_1 = \frac{1}{15000}\left(0 - 5000v_1 + 5000u_2 + 5000v_2 + 10000u_3 - 0v_3\right) = 0.1143 \text{ mm}$$

The reaction forces at the points of restraint can be recovered from the global equation.

$$X_4 = 0u_1 + 0v_1 - 86950u_2 + 23085v_2 - 0u_3 - 50018v_3 = 2000 \text{ N}$$

$$Y_4 = 0u_1 + 0v_1 + 26933u_2 - 26933v_2 - 50018u_3 - 0v_3 = -84 \text{ N}$$

$$X_5 = 0u_1 + 0v_1 + 0u_2 + 0v_2 - 86950u_3 + 26933v_3 = -2000 \text{ N}$$

$$Y_5 = 0u_1 + 0v_1 + 0u_2 + 0v_2 + 23085u_3 - 26933v_3 = 1084 \text{ N}$$

The strains in the shear web can be recovered from Equation (4.7), and the stresses from Equations (4.3) and (4.4a).

For element 5:

$$\begin{bmatrix} \varepsilon_x \\ \varepsilon_y \\ \gamma_{xy} \end{bmatrix} = [B]\{u\} = \frac{1}{2.25 \times 10^6} \begin{bmatrix} -1500 & 0 & 1500 & 0 & 0 & 0 \\ 0 & 0 & 0 & -1500 & 0 & 1500 \\ 0 & -1500 & -1500 & 1500 & 1500 & 0 \end{bmatrix} \begin{bmatrix} 0 \\ 0 \\ -0.0179 \\ -0.0414 \\ 0.0143 \\ -0.0280 \end{bmatrix} = 1\times10^{-6} \begin{bmatrix} -11.9 \\ 8.9 \\ -6.1 \end{bmatrix}$$

$$\begin{bmatrix} \sigma_x \\ \sigma_y \\ \tau_{xy} \end{bmatrix} = [D]\begin{bmatrix} \varepsilon_x \\ \varepsilon_y \\ \gamma_{xy} \end{bmatrix} = 76923 \times 1\times10^{-6} \begin{bmatrix} 1 & 0.3 & 0 \\ 0.3 & 1 & 0 \\ 0 & 0 & 0.35 \end{bmatrix} \begin{bmatrix} -11.9 \\ 8.9 \\ -6.1 \end{bmatrix} = \begin{bmatrix} -0.71 \\ 0.41 \\ -0.16 \end{bmatrix}$$

For element 6:

$$\begin{bmatrix} \varepsilon_x \\ \varepsilon_y \\ \gamma_{xy} \end{bmatrix} = [B]\{u\} = \frac{1}{2.25 \times 10^6} \begin{bmatrix} 1500 & 0 & -1500 & 0 & 0 & 0 \\ 0 & 0 & 0 & 1500 & 0 & -1500 \\ 0 & 1500 & 1500 & -1500 & -1500 & 0 \end{bmatrix} \begin{bmatrix} 0.0143 \\ -0.0280 \\ 0 \\ 0 \\ 0 \\ 0 \end{bmatrix} = 1\times10^{-6} \begin{bmatrix} 9.5 \\ 0 \\ -18.7 \end{bmatrix}$$

$$\begin{bmatrix} \sigma_x \\ \sigma_y \\ \tau_{xy} \end{bmatrix} = [D] \begin{bmatrix} \varepsilon_x \\ \varepsilon_y \\ \gamma_{xy} \end{bmatrix} = 76923 \times 1 \times 10^{-6} \begin{bmatrix} 1 & 0.3 & 0 \\ 0.3 & 1 & 0 \\ 0 & 0 & 0.35 \end{bmatrix} \begin{bmatrix} 9.5 \\ 0 \\ -18.7 \end{bmatrix} = \begin{bmatrix} 0.73 \\ 0.22 \\ -0.50 \end{bmatrix}$$

Consider now the distribution of shear stress that is implied by this analysis. It has been demonstrated that the shear stress in each triangular member is constant. The predicted stress distribution is therefore that illustrated below.

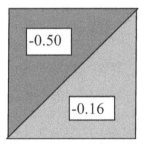

Clearly there is a discontinuity of stress at the intersection between the elements. An alternative presentation of this stress distribution is to assign the element stresses to the nodes and to average stresses from adjacent elements that meet at a node. Assuming then that the stress varies linearly between the nodes, the distribution of stress in the web might then be represented as illustrated below.

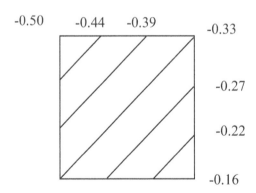

This latter representation looks more realistic but it should be recognised that it has been generated from the first one simply by an averaging process in the post-processing.

The problem has been modelled using finite element programme.

The structure has been modelled and analysed with two triangular elements representing the shear web. Three alternative models are illustrated below, together with the original. The first alternative has the triangular elements orientated differently, and the second and third feature one and one hundred quadrilateral elements, respectively. The results for each are tabulated.

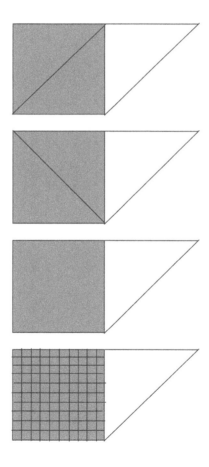

v (1)	−0·3736 mm	−0·3584 mm	−0·4234 mm	−0·5641 mm
Y (4)	−84 N	1148 N	356 N	475 N
Y (5)	1084 N	−148 N	644 N	525 N

Distribution of shear stress in web

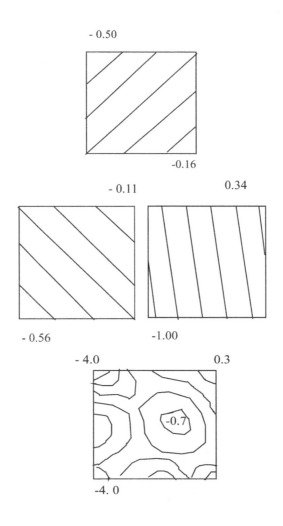

The displacement-formulation finite element analysis will fundamentally converge from below. This is discussed in more detail later, but it means that the displacements will always be underestimated, provided the element formulation conforms to certain rules. The tip displacement increases as the model is improved, and that which was predicted by the last model is close to the true value. Even the very crude models have yielded answers that are of the right order of magnitude, and this is all that such models are good for.

The computed vertical reactions vary widely between the four models. The finest mesh indicates that the reactions at the two nodes are almost equal, and this seems intuitively reasonable. The triangular meshes give very poor results.

The distribution of shear stress in the web was expected to be complex, and indeed the finest mesh shows that this is the case. The peak value in the centre of the web, approximately -0.7 N/mm^2, is a good approximation to the true value. The values at the corners of the panel, where concentrated loads are introduced, must be interpreted carefully. The analyst should note the magnitudes of the stresses on the free edges of the panel. Some components of stress must be zero at the free edge, and this might be used as one indicator of the likely accuracy of the analysis results. The issues of model verification and results interpretation will be discussed in some detail later.

It is clear from this numerical study, that the results of analyses featuring few elements in areas of rapidly changing stress are of very little value. The analyst must ensure that there are sufficiently many elements in a mesh to model the true stress distribution with the required accuracy.

4.6 Restrictions on Element Formulations—Completeness and Compatibility

The concept of shape functions was introduced in Section 3.2, and shape functions representing the linear variation of displacement along a bar were developed. The idea was extended in Sections 4.3.1 and 4.4.1 in order to develop shape functions for triangular and quadrilateral continuum elements also featuring a linear variation of displacement. In all cases, the choice of shape function was intuitively clear. However, the functions chosen had some important properties that will now be discussed in more detail.

4.6.1 *Attributes of the polynomial form of shape function—number of terms and differentiability*

The first observation about the shape functions chosen so far is that all are polynomial in form, and that the number of terms in each polynomial is equal to the number of degrees of freedom that are associated with the definition of the shape functions. Thus, the shape function for the bar element has two terms because it is required to have two independent nodal displacements in the element, and the shape functions for the triangular and quadrilateral elements have three and four terms respectively (see Equations 4.14 and 4.20).

The principle of virtual displacements, used to impose the equilibrium condition, contains terms that are functions of the first derivatives with respect to the spatial co-ordinates of the displacement functions. It is important, therefore, that the shape functions chosen are at least once differentiable over the element. The polynomial representations chosen satisfy this condition.

4.6.2 *Completeness—constant strain and rigid body modes*

A body can move from its original position to a new location without deformation and, therefore, without any stress or strain being generated within it. It would not be appropriate, therefore, to choose a shape function that did not permit such rigid body movement. For the one-dimensional bar element, for example, both nodes could translate by the same amount, u_0, and no strain would be imposed on the bar. This implies that one of the terms in the shape function must be a constant so that the bar could translate with a zero value of strain at all points within the bar.

In addition to the rigid body modes, the body must also be able to adopt a condition of constant strain. This is intuitively reasonable because as more and more elements are used to model a structure, the strain in each element must approach a constant value along the element. This implies that the linear terms are required in the polynomials, so that the derivative of the displacement can be constant on the element.

By definition, an element that has displacement functions that can represent the rigid body displacements and the constant strain states is said to be complete.

It is noted that the polynomial forms with the minimum number of terms of the shape functions for the two-node bar and for the three-node triangle are prescribed by the requirement for completeness.

4.6.3 Compatibility

The shape functions determine the variation of displacement within the element. The concept of compatibility means that the displacements across the element boundaries are continuous. This ensures that no gaps can appear between the elements.

Compatibility is automatically assured between two-node bar elements because they join only at the nodes. The compatibility between three-node triangular elements can be illustrated using the shear web modelled in Section 4.5. Clearly, it is necessary for the displacements along the common edge of elements 5 and 6 to be compatible. Substituting the co-ordinates and areas of the two triangular elements into Equation (4.14) and the result into Equation (4.12a) yields equations for the x direction displacement of each element.

Element (5):

$$u = \frac{1}{1500^2}\left[\left(1500^2 - 1500x + 0y\right)u_4 + \left(0 + 1500x - 1500y\right)u_2 + \left(0 + 0x + 1500y\right)u_3\right]$$

Element (6):

$$u = \frac{1}{1500^2}\left[\left(0 + 1500x + 0y\right)u_3 + \left(0 - 1500x + 1500y\right)u_5 + \left(1500^2 + 0x - 1500y\right)u_4\right]$$

On the common boundary between the elements (edge 3–4), the equation of the line is $x = y$.

On $x = y = \xi$,

$$u_{element\ 5} = \frac{1}{1500^2}\left[\left(1500^2 - 1500\xi\right)u_4 + \left(0\right)u_2 + \left(1500\xi\right)u_3\right] = u_{element\ 6}$$

And, hence, the x-direction displacement along the common edge is the same in both elements. Using exactly the same method, it can be shown that the y-direction displacements are also compatible. Note that, although the displacements are compatible, the strains and the stresses in the two elements are not necessarily so, as demonstrated numerically in Section 4.5.

4.6.4 Conforming elements

Elements that are both complete and compatible are referred to as *conforming*. It is not difficult to satisfy these requirements in continuum elements, but the requirements of compatibility in particular can be difficult to satisfy for plate and shell elements.

4.6.5 Convergence

It can be shown that if conforming elements are used, then the accuracy of a finite element analysis will increase as each element is broken down into smaller and smaller elements. The analysis is said to converge, and, furthermore, the convergence is monotonic. The displacements calculated in the structure will always be less than or equal to the true displacements.

4.6.6 Non-conforming elements and the patch test

Although monotonic convergence can only be guaranteed for conforming elements, there are several non-conforming elements (particularly plates and shells) in practical use in commercial finite element systems. The results of analyses performed using these elements are often sufficiently accurate for all practical purposes.

It can be shown that analyses of structures modelled using elements that violate the compatibility requirement will still converge, provided that the elements are capable of reproducing a constant strain condition when joined together in an assembly. This can be thought of as an extended completeness condition referring to an assembly of elements, rather than to one element in isolation. It is possible, for non-conforming elements, that assemblies of elements might be incapable of reproducing the constant strain condition even though the individual elements are complete. Non-conforming elements are, therefore, required to pass the *patch test*, in which a constant strain field is applied to an assembly of arbitrarily-orientated elements. If the element strains do represent the constant strain condition, then the test is passed. A typical mesh for the patch test of a two-dimensional element is illustrated in Figure 4.11.

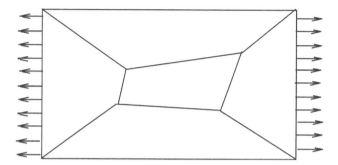

Figure 4.11 Patch test of a two-dimensional element.

Another part of the Patch Test is to check for strain-free rigid body motion, both translational and rotational. This entails applying a prescribed displacement to the external nodes of the patch of elements and checking to see that the internal nodes have moved by exactly the same amount and that there are no strains created in any of the elements. It is necessary to check each freedom independently, e.g., for a 2D element it is necessary to check that x and y translations and rotation about the z axis do not cause strains.

Chapter 5

Computational Implementation of the Finite Element Method

The finite element method is enormously versatile but requires a computer for any practical calculation. The tedium of solving even the simplest problem using pen, paper and a calculator has been amply demonstrated in earlier sections.

The stiffness matrix that represents any real structure is likely to be large, and sometimes very large indeed. It would not be uncommon for an aircraft model to contain over 100,000 degrees of freedom, giving a stiffness matrix with ten thousand million entries. Clearly then, the finite element programmer will be concerned with the development of efficient techniques for storage of the stiffness matrix and of efficient algorithms for solution of the resulting equations.

This chapter is intended to introduce the use of finite element analysis, rather than its coding, but some knowledge of programming and solution methodology is required for efficient use of the software. The purpose of this section is to discuss some of the issues associated with the computational implementation of the finite element method.

5.1 Solution Methodologies—Frontal v Banded

A linear static finite element analysis requires the solution of an equation of the form:

$\{F\} = [k] \{u\},$

where $\{F\}$ is a vector representing the (usually known) forces applied at the nodes and $\{u\}$ is a vector representing the (usually unknown) displacements at the nodes. The stiffness matrix of the structure, $[k]$, is assembled by the system based on the types of element specified and on their geometry and material properties. Since stiffness terms will only occur in the matrix where degrees of freedom are connected by an element, there will be a large number of zeros in the stiffness matrix of a typical structure. This is illustrated by the finite element model of the flat membrane shown below.

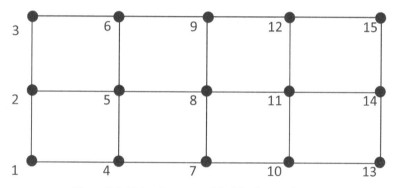

Figure 5.1 Finite element model of the flat membrane.

There are fifteen nodes in the structure and two degrees of freedom per node (translations in the x and y directions). The stiffness matrix therefore has dimensions of 30×30. The first column of the matrix contains all of those entries from degrees of freedom that are in the same element or elements as node 1. There are four such nodes (including node 1 itself) and, therefore, a total of eight non-zero entries (two degrees of freedom per node). A node in the middle of the structure, such as node 8, is connected to 9 nodes including itself and there are, therefore, eighteen non-zero entries in the columns associated with that node. The stiffness matrix for this structure is illustrated diagrammatically overleaf.

```
x x x x 0 0 x x x x 0 0 0 0 0 0 0 0 0 0 0 0 0 0 0 0 0 0 0 0
x x x x 0 0 x x x x 0 0 0 0 0 0 0 0 0 0 0 0 0 0 0 0 0 0 0 0
x x x x x x x x x x x x 0 0 0 0 0 0 0 0 0 0 0 0 0 0 0 0 0 0
x x x x x x x x x x x x 0 0 0 0 0 0 0 0 0 0 0 0 0 0 0 0 0 0
0 0 x x x x 0 0 x x x x 0 0 0 0 0 0 0 0 0 0 0 0 0 0 0 0 0 0
0 0 x x x x 0 0 x x x x 0 0 0 0 0 0 0 0 0 0 0 0 0 0 0 0 0 0
x x x x 0 0 x x x x 0 0 x x x x 0 0 0 0 0 0 0 0 0 0 0 0 0 0
x x x x 0 0 x x x x 0 0 x x x x 0 0 0 0 0 0 0 0 0 0 0 0 0 0
x x x x x x x x x x x x x x x x 0 0 0 0 0 0 0 0 0 0 0 0 0 0
x x x x x x x x x x x x x x x x 0 0 0 0 0 0 0 0 0 0 0 0 0 0
0 0 x x x x 0 0 x x x x 0 0 x x x x 0 0 0 0 0 0 0 0 0 0 0 0
0 0 x x x x 0 0 x x x x 0 0 x x x x 0 0 0 0 0 0 0 0 0 0 0 0
0 0 0 0 0 0 x x x x 0 0 x x x x 0 0 x x x x 0 0 0 0 0 0 0 0
0 0 0 0 0 0 x x x x 0 0 x x x x 0 0 x x x x 0 0 0 0 0 0 0 0
0 0 0 0 0 0 x x x x x x x x x x x x x x x x 0 0 0 0 0 0 0 0
0 0 0 0 0 0 x x x x x x x x x x x x x x x x 0 0 0 0 0 0 0 0
0 0 0 0 0 0 0 0 x x x x 0 0 x x x x 0 0 x x x x 0 0 0 0 0 0
0 0 0 0 0 0 0 0 x x x x 0 0 x x x x 0 0 x x x x 0 0 0 0 0 0
0 0 0 0 0 0 0 0 0 0 0 0 x x x x 0 0 x x x x 0 0 x x x x 0 0
0 0 0 0 0 0 0 0 0 0 0 0 x x x x 0 0 x x x x 0 0 x x x x 0 0
0 0 0 0 0 0 0 0 0 0 0 0 x x x x x x x x x x x x x x x x x x
0 0 0 0 0 0 0 0 0 0 0 0 x x x x x x x x x x x x x x x x x x
0 0 0 0 0 0 0 0 0 0 0 0 0 0 x x x x 0 0 x x x x 0 0 x x x x
0 0 0 0 0 0 0 0 0 0 0 0 0 0 x x x x 0 0 x x x x 0 0 x x x x
0 0 0 0 0 0 0 0 0 0 0 0 0 0 0 0 0 0 x x x x 0 0 x x x x 0 0
0 0 0 0 0 0 0 0 0 0 0 0 0 0 0 0 0 0 x x x x 0 0 x x x x 0 0
0 0 0 0 0 0 0 0 0 0 0 0 0 0 0 0 0 0 x x x x x x x x x x x x
0 0 0 0 0 0 0 0 0 0 0 0 0 0 0 0 0 0 x x x x x x x x x x x x
0 0 0 0 0 0 0 0 0 0 0 0 0 0 0 0 0 0 0 0 x x x x 0 0 x x x x
0 0 0 0 0 0 0 0 0 0 0 0 0 0 0 0 0 0 0 0 x x x x 0 0 x x x x
```

This diagram illustrates several attributes of stiffness matrices.

- They are extremely tedious to construct and to manipulate and all such activity is best left to the computer.
- They are large even for very small structures, and any finite element solution will consume significant amounts of computer resources.
- They are heavily populated with zeros (sparse). This is a very small model, and it is clear that the stiffness matrix will become increasingly sparse as the size of the model is increased.
- The non-zero terms are clustered in a band around the main diagonal. The matrix is said to be banded.

5.1.1 Banded solver

The bandwidth of the stiffness matrix is defined as the maximum width of any row, where the width of a row is the number of terms in the row that must be included before all entries to left and right of the included terms are zero. For the matrix illustrated, the bandwidth is eighteen. The semi-bandwidth is approximately one half of the bandwidth, and is defined as the width of the row starting at the main diagonal and moving to the right to the point at which there are no more non-zero entries in the row. For the matrix illustrated, it is ten. In fact, the semi-bandwidth is equal to the maximum node number difference on any element plus one multiplied by the number of degrees of freedom per node. In this case the difference between the highest and lowest numbered nodes on every element is four, and the semi-bandwidth is $(4 + 1) \times 2 = 10$.

A matrix that contains a large number of zeros is referred to as a sparse matrix, and there are several algorithms based on Gaussian Elimination that deal efficiently with matrices of this type. In principle, the time taken for a solution of a finite element equation of the form described tends to be proportional to the square of the semi-bandwidth, and so it is very important to number the structure in such a way that this is minimised. If the nodes in the mesh illustrated in the above example were renumbered in bands from left to right, rather than bottom to top, the semi-bandwidth would be increased to fourteen. The cost of solution would, therefore, almost double. In principle, it is best to number firstly in the direction that contains the fewest nodes.

5.1.2 *Frontal solver*

The frontwidth of a solution is a very similar concept to that of bandwidth. In some finite element analysis systems, the progress of the solution is determined by the ordering of the elements, rather than by the ordering of the nodes, and degrees of freedom are eliminated from the equation once all elements that reference them have been processed. This has the effect of reducing (often dramatically) the size of the stiffness matrix that has to be stored at any one time. The degrees of freedom that are currently contributing to the stiffness matrix are referred to as the active degrees of freedom. The frontwidth (or wavefront) of the solution is the number of degrees of freedom that are active at any one time.

For the model discussed previously, the frontal solution would commence by assembling all of the stiffness terms for element 1. The program would then note that no more elements are connected to node 1, and so the stiffness matrix would be reduced by eliminating all of the terms connected with that node, expressing them in terms of degrees of freedom that remain. The stiffness terms from element 2 would then be assembled into the stiffness matrix, and then the terms associated with nodes 2 and 3 would be eliminated. The stiffness terms from element 3 would then be assembled into the matrix, expanding it to include the terms associated with nodes 4 and 5. The solution proceeds in this manner until finally only those terms associated with the stiffness of the last element remain. The displacements in this element are solved, and then the program sweeps back through the eliminated freedoms, calculating their magnitudes from those of the freedoms already evaluated. The back-substitution step is a feature of both frontal and banded solvers.

The effect in the frontal solution is that of a wave of active degrees of freedom moving through the finite element model until a solution is obtained for the last one, and then sweeping back evaluating all other freedoms. This is illustrated below. The frontal solver is sometimes called a wavefront solver for this reason. The numbering should commence in the direction in which there are the fewest elements.

The commericially available FEM programmes usually use a frontal solver, so the student should pay attention to the element numbering scheme when constructing models. The system does offer a frontwidth optimisation algorithm, but good initial numbering is likely to produce the best results.

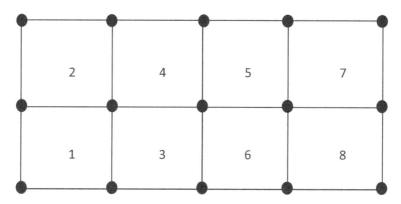

Figure 5.2 Elements numbering.

5.2 Storage of the Stiffness Matrix

Some of the attributes of the stiffness matrix were discussed in Chapter 2 when the matrix stiffness method was first introduced. Two additional important attributes were noted when stiffness matrix for a larger model was written down. The stiffness matrix is:

- sparse   (Most of the entries are zeros.)
- banded   (The non-zero entries are clustered around the main diagonal.)

The stiffness matrix is sparse because it contains non-zero terms only where the respective row and column degrees of freedom are directly connected by an element. In a general continuum a node will be directly connected to only a few other nodes, and therefore most of the entries in any row or column will be zero. The matrix will be banded if the nodes are numbered in such a way that adjacent nodes are numbered consecutively.

The attributes of the stiffness matrix discussed above suggest that techniques might be available to store it efficiently.

The optimal form of storage is, of course, dependent on the solution algorithm. As discussed in the previous chapter, one of the most important solution algorithms, the wavefront method, reduces storage requirements by never actually assembling the complete stiffness matrix.

5.3 Numerical Integration—Gaussian Quadrature

The stiffness matrix of an element is defined in its formulation as an integral of some quantities over the space occupied by the element. The computer must have some method of calculating the magnitude of this integral. It has been demonstrated that closed-form integration of the terms to form the stiffness matrix is impractical for all but the simplest elements. The alternative is to employ a numerical integration scheme. There are many numerical integration algorithms, but the one that has found almost universal favour in the coding of the finite element method is that based on Gaussian quadrature.

The advantages of the parametric formulation for the development of element stiffness matrices will be demonstrated in Chapter 7. For these elements, the integration limits are always −1 and +1, because integration takes place with respect to the parametric co-ordinates. The principles of Gaussian integration are illustrated below with particular regard to integration in parametric space. Consider the integration of a function defined by a straight line over this interval.

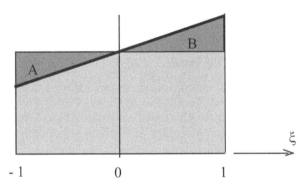

Figure 5.3 Principle of Gaussian integration.

Consider the integral of the function defined by the straight line illustrated above. The integral is just the area between the line and the ξ axis over the interval shown. However, because the areas of the triangles A and B are equal, this integral can be calculated simply by calculating the value of the function at $\xi = 0$ and multiplying by two (the length of the ξ axis). The point $\xi = 0$ would be called the *Gauss Point*, and the multiplying factor (two) would be called the *Gauss Weight*. The integration has been achieved, and is exact, using single point Gaussian integration.

This simple principle can be extended to the evaluation of the integrals of polynomial functions of a higher order. The locations of the Gauss points and the associated Gauss weights for particular orders of integration can be incorporated into a computer program. For the quadratic or the cubic curves illustrated below, two-point Gaussian integration would yield the exact value for the integral. In this case the Gauss points are located at $\xi = \pm 1/\sqrt{3}$ on the interval and the Gauss weight is one at each point.

It can be shown that *n*-point Gaussian integration, with an appropriate choice of weights, yields an exact solution for the integration of a polynomial of order $(2n-1)$.

It was shown in Chapter 6 that the Jacobian matrix for a four node quadrilateral with sides parallel to the global axes is constant over the element. The terms in the stiffness matrix can be expressed as integrals of quadratics in ξ and η. This means that two-point Gaussian integration in each of the parametric directions will yield an exact stiffness matrix for this element. Because integration in two directions is required, there are four Gauss points in the element and the integration is usually referred to as *two x two*. The locations of the (Gauss) integration points are illustrated below.

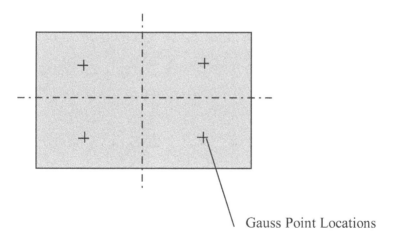

Gauss Point Locations

Figure 5.4 Gaussian integration points.

The situation is rather more complex when the element is distorted, causing the Jacobian and its inverse to vary over the element. In this case the two x two integration will no longer be exact, and higher order integration might be indicated. The use of 2×2 integration will yield lower values of the stiffness terms than would exact integration. However, it is a fundamental property of the displacement-based elements that the stiffness of the structure will be over-predicted. In practice, the errors produced by the lower order integration tend to compensate for the error inherent in the formulation and so-called *reduced integration* can actually improve the performance of the element.

This idea can be developed further, and it is possible to select different orders of integration for different components of the stress and strain components that contribute to the element stiffness. Some components, particularly those from the shear terms, are often under-integrated quite deliberately so as to improve the performance of the element.

Reduced integration can have its drawbacks, and it can be shown that it is possible for the eight-node quadrilateral to be strained into an hourglass shape whilst the stresses at the four low-order integration Gauss points are all zero. This means that there is a displaced configuration of the element that will give rise to no strain energy (zero-energy mode) if reduced integration is adopted. This only occurs with a single element and a mesh of two elements or more will not suffer from this problem.

Chapter 6

Beams, Plates, Shells and Solids

The one-dimensional bar and two-dimensional membrane elements can be used to model a wide range of engineering structures. Inevitably though, the analyst will soon encounter problems for which these elements are inappropriate. In conventional analysis, the bending theory is used to investigate structures that are long relative to their cross-sectional dimensions. The fundamental methodology used to develop this theory can be applied to develop plate and shell theories to study the behaviour of structures formed from thin sheet material. Finite elements can also be developed to model beam, plate and shell structures. Some of the more complex elements that will be found in the library of any commercial finite element programme are briefly described in this chapter.

6.1 Solid Elements

It is, of course, possible to extend the two-dimensional membrane theory into three dimensions to develop an element capable of representing a three-dimensional continuum. These elements are referred to as solid elements. When the solid is a hexahedron it is often called a brick element. The simplest form of brick element has eight nodes, one at each vertex, with three degrees of freedom representing three mutually orthogonal components of displacement at each node, as in Figure 6.1(a).

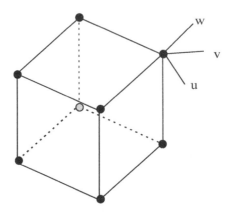

a) Eight-node Brick Element

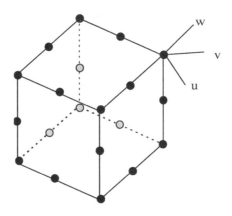

b) Twenty-node Brick Element

Figure 6.1 Brick element.

The element stiffness is represented by a 24 × 24 matrix. This implies that there are 576 terms in the stiffness matrix of one element, and it should be remembered that this element is capable of modelling little more than a constant stress field.

If mid-side nodes are introduced to capture linear stress variations (as in Figure 6.1(b)), the element then has 20 nodes, 60 degrees of freedom, and 3600 entries in its stiffness matrix. Clearly, three-dimensional models constructed from solid elements are going to be very expensive in computer resources, in terms of both storage requirements and run times.

There are some structures for which solid elements offer the only route to an accurate analysis model, but in many cases it is possible to make some assumptions about the behaviour of the structure that allow the use of a less computationally-expensive element.

There is a temptation for the inexperienced analyst to use solid elements excessively, including in situations where other elements should be preferred. It should be noted that a model consisting of solid elements in which the aspect ratios are large might be prone to ill-conditioning due to the high ratio of in-plane to through-thickness stiffness.

6.2 A Beam Element

It has been shown in Section 4.4.3 that the two-dimensional membrane element can be used to model a beam subjected to a bending moment. However, many four-node quadrilaterals are needed in order to approximate the true distribution of lateral displacement as a series of straight lines. Furthermore, much time is spent in the calculation of stiffness terms associated with the transverse stretching and shear deformation of the elements, and these are neglected in the bending. It would be much more efficient to develop an element that is capable of a much higher resolution of the lateral deflections, but that did not waste effort on the unimportant stretching and shear terms.

6.2.1 Nodal degrees of freedom

A beam that is subjected to a series of discrete bending moments and shear forces deflects into a cubic shape. Only the lateral deflection is considered important, and a four-node element could be defined with a single degree of freedom at each node.

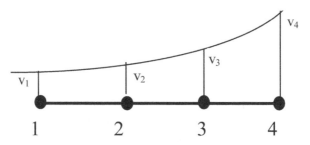

Figure 6.2 Nodal displacements.

The deflection at any point along the beam could then be defined in terms of these nodal displacements, and then the stiffness matrix could be evaluated by appealing to the principle of virtual dispacements. This beam would require the definition of the co-ordinates of each of the four nodes.

In practice, a straight beam is defined in terms of the co-ordinates of its two end nodes, so the beam element features only two nodes. Four degrees of freedom are required to model the cubic displacement field and the displacements of the two intermediate nodes are replaced by the rotation of the beam at the end nodes.

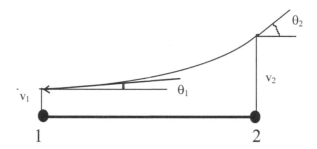

Figure 6.3 Nodal rotations.

The introduction of the nodal rotations as degrees of freedom of the element is a very important concept. These additional degrees of freedom at the nodes constitute the main difference between the continuum elements, in which the displacement of a node is defined entirely in terms of its new geometric location, and the beam, plate and shell elements.

Mixed models featuring both types of element should be treated with caution, because the elements are fundamentally incompatible.

6.2.2 Shape functions

Consider the beam illustrated below, subjected to end forces and moments.

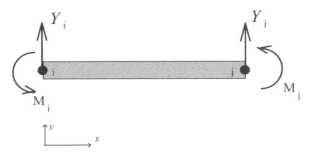

Figure 6.4 Beam with end forces and moments.

There are two degrees of freedom at each node, the transverse deflection v and the rotation θ. Equations are required that relate the force and moments at the nodes (Y_i, M_i, Y_j, M_j) to the displacements and rotations there. Unlike the two-dimensional bar element, in which the degrees of freedom at the nodes are independent of one another, the degrees of freedom on the beam element are related by the equation:

$$\theta = \frac{dv}{dx} \tag{6.1}$$

A shape function is required that can describe the distribution of the nodal variables at all points along the beam. The fundamental variable on the beam is the transverse displacement v, and the simplest shape function is in the form of a polynomial. In this case, four terms in the polynomial will be required to satisfy the four boundary conditions.

$$v = a_0 + a_1 x + a_2 x^2 + a_3 x^3 \tag{6.2}$$

Then from Equation (6.1),

$$\theta = a_1 + 2a_2x + 3a_3x^2 \tag{6.3}$$

For simplicity it will be assumed that node i lies on the y axis ($x = 0$) and that the beam is of length L. The final result for the stiffness matrix is independent of this assumption.

Substituting the boundary conditions into Equations (6.1) and (6.2):

$$v_i = a_0$$

$$v_j = a_0 + a_1L + a_2L^2 + a_3L^3$$

$$\theta_i = a_1$$

$$\theta_j = a_1 + 2a_2L + 3a_3L^3$$

These equations can be solved for the unknown coefficients a.

$$a_0 = v_i$$

$$a_1 = \theta_i L^3$$

$$a_2 = -\frac{3}{L^2}\left(v_i - v_j\right) - \frac{1}{L}\left(2\theta_i - \theta_j\right)$$

$$a_3 = \frac{2}{L^3}\left(v_i - v_j\right) + \frac{1}{L^2}\left(\theta_i + \theta_j\right)$$

Then:

$$v = \begin{bmatrix} N_1 & N_2 & N_3 & N_4 \end{bmatrix} \begin{bmatrix} v_i \\ \theta_i \\ v_j \\ \theta_j \end{bmatrix}, \tag{6.4}$$

where:

$$N_1 = 1 - \frac{3x^2}{L^2} + \frac{2x^3}{L^3}$$

$$N_2 = x - \frac{x^2}{L} + \frac{x^3}{L^2}$$

$$N_3 = \frac{3x^2}{L^2} - \frac{2x^3}{L^3}$$

$$N_4 = -\frac{x^2}{L} + \frac{x^3}{L^2}$$

It can readily be shown that Equation (6.4) yields the required values of the nodal degrees of freedom.

Node	N_1	N_2	N_3	N_4	$\partial N_1/\partial x$	$\partial N_2/\partial x$	$\partial N_3/\partial x$	$\partial N_4/\partial x$
1	1	0	0	0	0	1	0	0
2	0	0	1	0	0	0	0	1

6.2.3 *The stiffness matrix*

The stiffness matrix for the beam element of constant bending stiffness can be derived using the principle of virtual displacements, Equation (3.2). The final result is:

$$\begin{bmatrix} Y_i \\ M_i \\ Y_j \\ M_j \end{bmatrix} = \frac{EI}{L^3} \begin{bmatrix} 12 & 6L & -12 & 6L \\ 6L & 4L^2 & -6L & 2L^2 \\ -12 & -6L & 12 & -6L \\ 6L & 2L^2 & -6L & 4L^2 \end{bmatrix} \begin{bmatrix} v_i \\ \theta_i \\ v_j \\ \theta_j \end{bmatrix}$$

or

$$\{F\} = [k]\{u\}. \tag{6.5}$$

This element will be capable of modelling any beam structure with discrete lateral loads and moments. The displacement results will be exact according to the bending theory. It should be noted that it is possible to combine the axial freedom associated with the bar element in order to develop a more general element, although bending and axial stretching will normally remain uncoupled.

6.2.4 Numerical example 1

The numerical solution of beam problems proceeds in exactly the same way as that of the bar problems that have already been considered.

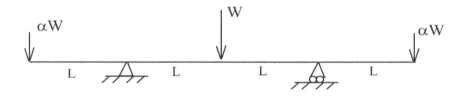

A beam of uniform bending stiffness EI is loaded and supported as illustrated above. Calculate the value of the constant α for which the vertical displacement at the ends of the beam will be zero.

The beam is symmetrical about its mid-point, and so it is only necessary to model one-half of the structure.

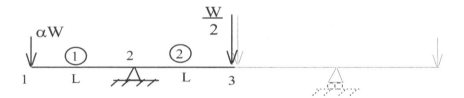

The stiffness matrices for each of the two beam elements are written below.

Element 1: the beam is of length L and of bending stiffness EI.

$$
\begin{bmatrix} Y_1 \\ M_1 \\ Y_2 \\ M_2 \end{bmatrix}_1 = \frac{EI}{L^3} \begin{bmatrix} 12 & 6L & -12 & 6L \\ 6L & 4L^2 & -6L & 2L^2 \\ -12 & -6L & 12 & -6L \\ 6L & 2L^2 & -6L & 4L^2 \end{bmatrix} \begin{bmatrix} v_1 \\ \theta_1 \\ v_2 \\ \theta_2 \end{bmatrix}
$$

Element 2: the beam is of length L and of bending stiffness EI.

$$
\begin{bmatrix} Y_2 \\ M_2 \\ Y_3 \\ M_3 \end{bmatrix}_2 = \frac{EI}{L^3} \begin{bmatrix} 12 & 6L & -12 & 6L \\ 6L & 4L^2 & -6L & 2L^2 \\ -12 & -6L & 12 & -6L \\ 6L & 2L^2 & -6L & 4L^2 \end{bmatrix} \begin{bmatrix} v_2 \\ \theta_2 \\ v_3 \\ \theta_3 \end{bmatrix}
$$

The element stiffness matrices are assembled into a global stiffness matrix for the structure.

$$
\begin{bmatrix} Y_1 \\ M_1 \\ Y_2 \\ M_2 \\ Y_3 \\ M_3 \end{bmatrix} = \frac{EI}{L^3} \begin{bmatrix} 12 & 6L & -12 & 6L & 0 & 0 \\ 6L & 4L^2 & -6L & 2L^2 & 0 & 0 \\ -12 & -6L & 12+12 & -6L+6L & -12 & 6L \\ 6L & 2L^2 & -6L+6L & 4L^2+4L^2 & -6L & 2L^2 \\ 0 & 0 & -12 & -6L & 12 & -6L \\ 0 & 0 & 6L & 2L^2 & -6L & 4L^2 \end{bmatrix} \begin{bmatrix} v_1 \\ \theta_1 \\ v_2 \\ \theta_2 \\ v_3 \\ \theta_3 \end{bmatrix}
$$

Note once again that there are as many contributors to each of the entries on the main diagonal as there are members joining at the node under consideration. Adding the contributions together, the global stiffness matrix is calculated. At this stage, the known values of force and displacement at the nodes are entered into the force and displacement vectors.

$$
\begin{bmatrix}
-\alpha\,W \\
0 \\
Y_2 \\
0 \\
-0.5W \\
M_3
\end{bmatrix}
=
\frac{EI}{L^3}
\begin{bmatrix}
12 & 6L & -12 & 6L & 0 & 0 \\
6L & 4L^2 & -6L & 2L^2 & 0 & 0 \\
-12 & -6L & 24 & 0 & -12 & 6L \\
6L & 2L^2 & 0 & 8L^2 & -6L & 2L^2 \\
0 & 0 & -12 & -6L & 12 & -6L \\
0 & 0 & 6L & 2L^2 & -6L & 4L^2
\end{bmatrix}
\begin{bmatrix}
0 \\
\theta_1 \\
0 \\
\theta_2 \\
v_3 \\
0
\end{bmatrix}
$$

The rows and columns corresponding to the unknown forces and zero displacements are eliminated.

$$
\begin{bmatrix}
0 \\
0 \\
-0.5W
\end{bmatrix}
=
\frac{EI}{L^3}
\begin{bmatrix}
4L^2 & 2L^2 & 0 \\
2L^2 & 8L^2 & -6L \\
0 & -6L & 12
\end{bmatrix}
\begin{bmatrix}
\theta_1 \\
\theta_2 \\
v_3
\end{bmatrix}
$$

Reducing the stiffness matrix to upper triangular form using standard Gaussian elimination:

$$
\begin{bmatrix}
0 \\
0 \\
-0.5W
\end{bmatrix}
=
\frac{EI}{L^3}
\begin{bmatrix}
4L^2 & 2L^2 & 0 \\
0 & 7L^2 & -6L \\
0 & -6L & 12
\end{bmatrix}
\begin{bmatrix}
\theta_1 \\
\theta_2 \\
v_3
\end{bmatrix}.
$$

$$
\begin{bmatrix}
0 \\
0 \\
-0.5W
\end{bmatrix}
=
\frac{EI}{L^3}
\begin{bmatrix}
4L^2 & 2L^2 & 0 \\
0 & 7L^2 & -6L \\
0 & 0 & 48/7
\end{bmatrix}
\begin{bmatrix}
\theta_1 \\
\theta_2 \\
v_3
\end{bmatrix}
$$

The solution for the displacements can be determined by back-substitution:

$$
v_3 = \frac{-7W}{48 \times 2} \cdot \frac{L^3}{EI} = \frac{-7WL^3}{96EI}
$$

$$
\theta_2 = \frac{1}{7L^2}\left(0 \cdot \frac{L^3}{EI} + 6Lv_3\right) = \frac{-6WL^2}{96EI}
$$

$$
\theta_1 = \frac{1}{4L^2}\left(0 \cdot \frac{L^3}{EI} - 2L^2\theta_2 + 0 \cdot v_3\right) = \frac{3WL^2}{96EI}
$$

The reaction forces at the points of restraint can now be recovered from the global equation. In this case, only the first equation is of interest.

$$-\alpha\,W = \frac{EI}{L^3}\left(6L.\theta_1 + 6L.\theta_2 + 0 \cdot v_3\right)$$

Substituting for the displacements and rotations,

$$-\alpha\,W = \frac{EI}{L^3} \times \frac{WL^2}{96EI}\left(18L - 36L\right) \qquad \Rightarrow \qquad \alpha = \frac{18}{96} = \frac{3}{16} = 0.1875$$

6.2.5 Numerical example 2

Calculate the deflection and the internal bending moment at the point of application of the load in the fully built-in beam shown below.

Also calculate the reactions (forces and moments) that the walls exert on the beam.

$E = 200$ kN/mm² and $I = 4 \times 10^{-6}$ m⁴.

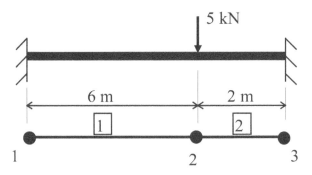

First calculate the stiffness matrices of the two elements using Equation (6.5).

Element 1 $L_1 = 6$ m and $EI = 0.8 \times 10^6$ Nm².

$$\begin{bmatrix} Y_1 \\ M_1 \\ Y_2 \\ M_2 \end{bmatrix}_1 = \frac{0.8 \times 10^6}{6^3} \begin{bmatrix} 12 & 36 & -12 & 36 \\ 36 & 144 & -36 & 72 \\ -12 & -36 & 12 & -36 \\ 36 & 72 & -36 & 144 \end{bmatrix} \begin{bmatrix} v_1 \\ \theta_1 \\ v_2 \\ \theta_2 \end{bmatrix}$$

Element 2 L_2 = 2 m and EI = 0.8 × 10^6 Nm^2.

$$\begin{bmatrix} Y_2 \\ M_2 \\ Y_3 \\ M_3 \end{bmatrix}_2 = \frac{0.8 \times 10^6}{2^3} \begin{bmatrix} 12 & 12 & -12 & 12 \\ 12 & 16 & -12 & 8 \\ -12 & -12 & 12 & -12 \\ 12 & 8 & -12 & 16 \end{bmatrix} \begin{bmatrix} v_2 \\ \theta_2 \\ v_3 \\ \theta_3 \end{bmatrix}$$

The element stiffness matrices are now assembled into a global stiffness matrix for the structure and the boundary conditions substituted into the force and displacement vectors:

$$\begin{bmatrix} Y_1 \\ M_1 \\ -5000 \\ 0 \\ Y_3 \\ M_3 \end{bmatrix} = 10^6 \begin{bmatrix} 0.044 & 0.133 & -0.044 & 0.133 & 0 & 0 \\ 0.133 & 0.533 & -0.133 & 0.266 & 0 & 0 \\ -0.044 & -0.133 & 1.244 & 1.067 & -1.2 & 1.2 \\ 0.133 & 0.266 & 1.067 & 2.133 & -1.2 & 0.8 \\ 0 & 0 & -1.2 & -1.2 & 1.2 & -1.2 \\ 0 & 0 & 1.2 & 0.8 & -1.2 & 1.6 \end{bmatrix} \begin{bmatrix} 0 \\ 0 \\ v_2 \\ \theta_2 \\ 0 \\ 0 \end{bmatrix}$$

The rows and columns corresponding to the known (zero) displacements are temporarily discarded.

$$\begin{bmatrix} -5000 \\ 0 \end{bmatrix} = 10^6 \begin{bmatrix} 1.244 & 1.067 \\ 1.067 & 2.133 \end{bmatrix} \begin{bmatrix} v_2 \\ \theta_2 \end{bmatrix}$$

Eliminating entries below the main diagonal in column 1:

$$\begin{bmatrix} -5000 \\ 4289 \end{bmatrix} = 10^6 \begin{bmatrix} 1.244 & 1.067 \\ 0 & 1.218 \end{bmatrix} \begin{bmatrix} v_2 \\ \theta_2 \end{bmatrix}$$

122

The solution for the displacements at the point of application of the load (i.e., node 2) can now be found by back-substitution:

$$\theta_2 = \frac{4289}{1.218 \times 10^6} = 0.00352 \text{ radians}$$

$$v_2 = \frac{\left(-5000 - 1.067 \times 10^6 \times \theta_2\right)}{1.244 \times 10^6} = -0.00704 \text{ m}$$

The previously discarded global equations are used to calculate the reaction forces that the built-in ends exert on the beam:

$$Y_1 = 10^6 \left\{ -0.044 \times \left(-0.00704\right) + 0.133 \times 0.00352 \right\} = 777.9 \text{ N}$$
$$M_1 = 10^6 \left\{ -0.133 \times \left(-0.00704\right) + 0.266 \times 0.00352 \right\} = 1872.6 \text{ Nm}$$
$$Y_3 = 10^6 \left\{ -1.2 \times \left(-0.00704\right) + \left(-1.2\right) \times 0.00352 \right\} = 4224 \text{ N}$$

$$M_3 = 10^6 \left\{ 1.2 \times \left(-0.00704\right) + 0.8 \times 0.00352 \right\} = -5632 \text{ Nm}$$

6.3 Plates and Shells

Plate and shell elements can be developed to represent the behaviour of thin-walled structures. The motivation for these elements is to save computational resources by reducing the number of degrees of freedom relative to the solid element. A further benefit is that, for any particular thin-walled structure, many fewer shell elements will be required than would be required for a full solid model due to constraints on element aspect ratios.

6.3.1 Plate elements

Plate elements can be thought of in terms of an extension of beam theory into two dimensions. Indeed the first attempts at such a theory were based on grillages of beams, with bending about two orthogonal axes. Such theories are restrictive, in that they neglect Poisson's ratio effects and also omit the effects of torsional moments generating shear stresses in the plate.

A more sophisticated theory, recognising the existence of the shear terms, is an extension of the beam theory, based on cubic lateral displacements along the generators in parametric space. The displacements in the x and y directions in a plane parallel to the neutral surface are assumed proportional to the distance from the neutral surface, and the strains at any point are calculated from membrane theory in the plane. The Poisson's ratio effects and the shear stress component are thus recognised. In practice, this element can handle out-of-plane bending moments reasonably well, but is not so good where torsional moments are present.

Derivatives of the element, such as one in which the shape functions associated with the element rotations are deliberately reduced from cubic to linear, can be used to improve performance in some circumstances. In common with other non-conforming elements, these elements compromise the bounded nature of the solution, hence, the need to carry out the Patch Test with such elements.

An alternative, and now very popular, approach to the development of a plate theory is the one represented by the Mindlin model. The fundamental principle in this approach is that the rotations at the nodes are independent of the transverse displacements. The rotations are not calculated by differentiation of w, but are regarded as separate quantities. This permits the plate to sustain transverse shear deformations, precluded by the more classical formulation. The Mindlin plate elements perform well in many practical situations. They are superior for the modelling of thick plates, because they recognise the transverse shear deformation and can be used successfully even for thin plates. For very thin plates, special consideration is given to the transverse shear terms in order to avoid numerical ill-conditioning.

6.3.2 Shell elements

In flat plate elements, membrane and bending actions are considered to be uncoupled, provided that the displacements are small. This is not true of curved shell elements, in which membrane loads on an edge will, in general, produce bending elsewhere. The formulation of shell elements is a very complex problem, and it is difficult to satisfy the traditional finite element criteria of completeness and compatibility. The field abounds with special elements, featuring clever tricks designed to improve element performance. Often reduced integration can be used to improve accuracy as well as to reduce computational effort. This can help to compensate for the inherent over-stiffness in displacement method finite elements.

The tricks employed in the formulation of shell elements mean that they should be used with greater care than is necessary for more conventional isoparametric plate elements. Because of the lack of conformity it is necessary to carry out verification checks (such as the patch test) on any new element of this type. Reduced integration can sometimes cause ill-conditioning, such as the hourglass mode, so this has to be guarded against in the model's construction.

Chapter 7

Parametric Element Formulations

Formulations for one and two-dimensional elements were developed in Chapters 3 and 4. The equations for the shape functions were specific to the elements because the terms in them were dependent on the geometrical co-ordinates of the nodes (see Equations (3.5), (4.14) and (4.20)). The development of an appropriate stiffness matrix for the bar element and for the constant strain triangle was not particularly onerous. The development of an appropriate stiffness matrix for the simplest rectangular element was shown, in contrast, to be relatively laborious, and it was suggested that a numerical integration technique should be employed to calculate each entry in the stiffness matrix. Further, the formulation was developed only for the rectangular element, and that for a general quadrilateral element, or even for a rectangular element skewed relative to the global axes, would be more difficult still. In order to overcome these difficulties, the parametric element formulations have been developed, in which the shape functions are defined in terms of natural co-ordinate systems (common to all elements of a particular type) and the elements are mapped onto real space using transformation matrices. This proves to be a very powerful technique, which permits the study of complicated geometries using relatively simple element formulations combined with numerical integration to calculate the stiffness matrices.

7.1 Isoparametric Bar Element

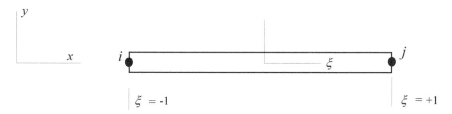

Figure 7.1 Isoparametric bar element.

7.1.1 Shape functions

As before, the displacement at all points in the element is expressed in terms of the displacements at the nodes.

$$u = N_i \, u_i + N_j \, u_j \tag{7.1}$$

The shape functions are now expressed in terms of the natural co-ordinate, ξ.

$$N_i = \frac{1}{2}(1 - \xi) \qquad : \qquad N_j = \frac{1}{2}(1 + \xi) \tag{7.2}$$

A mapping function is required that relates the global co-ordinates of the bar, x, to the natural co-ordinates, ξ. The most obvious such function is:

$$x = N_i \, x_i + N_j \, x_j. \tag{7.3}$$

In this case, the interpolation function for the element displacements, Equation (6.1), is the same as that for the co-ordinates, Equation (6.3). This is the basis of the *isoparametric* element formulation. If the interpolation function chosen for the co-ordinates is of a lower order than the one for the displacements then the elements are *subparametric*, and if it is of a higher order then the elements are *superparametric*. Isoparametric elements are predominant in commercial packages, and of the other two, subparametric elements have been found to be of greater use.

7.1.2 The element stiffness matrix

The element stiffness matrix is calculated based on Equation (3.15), and the [B] matrix is required from Equation (3.12).

$$[B] = \left[\frac{dN_i}{dx} \quad \frac{dN_j}{dx} \right] = \left[\frac{dN_i}{d\xi} \cdot \frac{d\xi}{dx} \quad \frac{dN_j}{d\xi} \cdot \frac{d\xi}{dx} \right]$$

Substituting Equations (7.2) into Equation (7.3):

$$x = \frac{x_i}{2}(1-\xi) + \frac{x_j}{2}(1+\xi)$$

Hence

$$\frac{dx}{d\xi} = -\frac{x_i}{2} + \frac{x_j}{2} = \frac{L}{2} \tag{7.4}$$

So

$$\frac{d\xi}{dx} = \frac{2}{L}$$

Differentiating Equations (7.2) wrt x and substituting the results plus the above into the expression for [B]:

$$[B] = \left[-\frac{1}{L} \quad \frac{1}{L} \right]. \tag{7.5}$$

The stiffness matrix is now found by integration (Equation (3.15)).

$$[k]_{elem} = \int_{x_i}^{x_j} [B]^T E[B] A_x \, dx = \int_{-1}^{1} [B]^T E[B] A_x \, J.d\xi$$

where J is the Jacobian relating the global co-ordinate system to the natural co-ordinate system.

$$J.d\xi = dx \qquad \text{So} \qquad J = \frac{dx}{d\xi} = \frac{L}{2} \tag{7.5.1}$$

128

For the bar of constant cross-section:

$$[k]_{elem} = AE\int_{-1}^{1}[B]^{T}[B] \; J \; d\xi = AE\int_{-1}^{1}\begin{bmatrix} -\dfrac{1}{L} \\ \dfrac{1}{L} \end{bmatrix}\begin{bmatrix} -\dfrac{1}{L} & \dfrac{1}{L} \end{bmatrix}\dfrac{L}{2}\; d\xi = \dfrac{AE}{L}\begin{bmatrix} 1 & -1 \\ -1 & 1 \end{bmatrix} \quad (7.6)$$

and the element stiffness matrix is the same as that calculated using the global co-ordinates, Equation (3.18).

7.2 Isoparametric Four-Node Quadrilateral Element

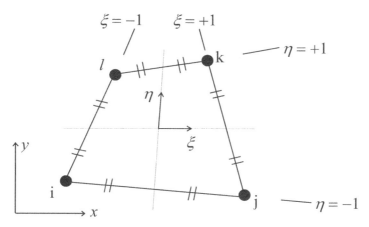

Figure 7.2 Isoparametric four-node quadrilateral element.

The force vector at node i is $\{X\}_i = \{X_i, Y_i\}^T$ and the displacement vector is $\{u\}_i = \{u_i, v_i\}^T$. Similar vectors apply at nodes j, k and l.

7.2.1 Shape functions

As before, assume that the displacements anywhere in the element can be expressed as functions of the displacements of the nodes. The functions must satisfy the boundary conditions, $u = u_i$ and $v = v_i$ at node i and similar conditions at nodes j, k and l. This requirement will be satisfied if functions of the form

$$u = N_i u_i + N_j u_j + N_k u_k + N_l u_l \qquad (7.7a)$$

$$v = N_i v_i + N_j v_j + N_k v_k + N_l v_l \qquad (7.7b)$$

are chosen, in which the shape function N_i adopts a value of zero at $(x = x_j, y = y_j)$, at $(x = x_k, y = y_k)$, and at $(x = x_l, y = y_l)$, and of one at $(x = x_i, y = y_i)$. Similar functions are chosen for the other shape functions. Again, the obvious functions, implying a linear variation of the displacement between the nodes, are chosen but this time the shape functions are defined in terms of the natural co-ordinates. A mapping function is required, relating points in parametric space to points in Cartesian space, and for the isoparametric formulation, the same shape functions are chosen.

$$x = N_i x_i + N_j x_j + N_k x_k + N_l x_l \qquad (7.8a)$$

$$y = N_i y_i + N_j y_j + N_k y_k + N_l y_l \qquad (7.8b)$$

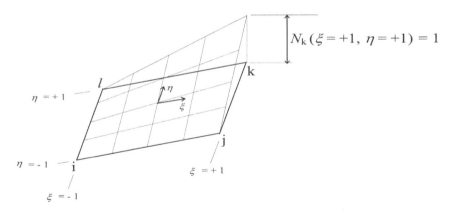

Figure 7.3 Isoparametric four-node quadrilateral element shape function.

$$N_I = \frac{1}{4}\big((1-\xi)(1+\eta)\big) \qquad : \qquad N_k = \frac{1}{4}\big((1+\xi)(1+\eta)\big)$$

$$N_i = \frac{1}{4}\big((1-\xi)(1-\eta)\big) \qquad : \qquad N_j = \frac{1}{4}\big((1+\xi)(1-\eta)\big)$$

$$. \ (7.9)$$

The form of these shape functions should be compared with those derived for the element in ordinary Cartesian co-ordinates. In contrast to that element, for which any element has its own shape functions depending on its position in the continuum, all elements of this type have common shape functions.

7.2.2 The element stiffness matrix

In order to write down the [B] matrix, the derivatives of the shape functions, with respect to the global co-ordinates, are required. By the chain rule of differentiation:

$$\frac{\partial}{\partial \xi} = \frac{\partial x}{\partial \xi}\cdot\frac{\partial}{\partial x} + \frac{\partial y}{\partial \xi}\cdot\frac{\partial}{\partial y} \qquad : \qquad \frac{\partial}{\partial \eta} = \frac{\partial x}{\partial \eta}\cdot\frac{\partial}{\partial x} + \frac{\partial y}{\partial \eta}\cdot\frac{\partial}{\partial y} \quad (7.10)$$

In matrix form:

$$\begin{bmatrix} \dfrac{\partial}{\partial \xi} \\[2mm] \dfrac{\partial}{\partial \eta} \end{bmatrix} = \begin{bmatrix} \dfrac{\partial x}{\partial \xi} & \dfrac{\partial y}{\partial \xi} \\[2mm] \dfrac{\partial x}{\partial \eta} & \dfrac{\partial y}{\partial \eta} \end{bmatrix} \begin{bmatrix} \dfrac{\partial}{\partial x} \\[2mm] \dfrac{\partial}{\partial y} \end{bmatrix} = [J]\begin{bmatrix} \dfrac{\partial}{\partial x} \\[2mm] \dfrac{\partial}{\partial y} \end{bmatrix}$$

where [J] is the Jacobian matrix for this transformation of co-ordinates. Noting the form of the shape functions, Equations (7.9), it can be seen that, in the general case, the first row of [J] contains terms that are linear in η and the second row contains terms that are linear in ξ.

$$[J] = \frac{1}{4}\begin{bmatrix} \big((-x_i + x_j + x_k - x_l) + (x_i - x_j + x_k - x_l)\,\eta\big) & \big((-y_i + y_j + y_k - y_l) + (y_i - y_j + y_k - y_l)\,\eta\big) \\[2mm] \big((-x_i - x_j + x_k + x_l) + (x_i - x_j + x_k - x_l)\,\xi\big) & \big((-y_i - y_j + y_k + y_l) + (y_i - y_j + y_k - y_l)\,\xi\big) \end{bmatrix}$$

$$(7.11)$$

The entries in the Jacobian vary from point to point in the element, depending on the slopes of the parametric lines in Cartesian space. One special case, that of a rectangular element with sides parallel to the global axes, is considered in the next section.

The derivatives of the displacements, with respect to the Cartesian co-ordinates, are required. These are calculated using the inverse of the Jacobian.

$$
\begin{bmatrix} \dfrac{\partial}{\partial x} \\[2mm] \dfrac{\partial}{\partial y} \end{bmatrix} = [J]^{-1} \begin{bmatrix} \dfrac{\partial}{\partial \xi} \\[2mm] \dfrac{\partial}{\partial \eta} \end{bmatrix}
\tag{7.12}
$$

Consider now the derivatives of u. Substituting Equations (7.7) into Equation (7.12):

$$
\begin{bmatrix} \dfrac{\partial u}{\partial x} \\[2mm] \dfrac{\partial u}{\partial y} \end{bmatrix} = [J]^{-1} \begin{bmatrix} \dfrac{\partial u}{\partial \xi} \\[2mm] \dfrac{\partial u}{\partial \eta} \end{bmatrix} = [J]^{-1} \begin{bmatrix} \dfrac{\partial N_i}{\partial \xi} & 0 & \dfrac{\partial N_j}{\partial \xi} & 0 & \dfrac{\partial N_k}{\partial \xi} & 0 & \dfrac{\partial N_l}{\partial \xi} & 0 \\[3mm] \dfrac{\partial N_i}{\partial \eta} & 0 & \dfrac{\partial N_j}{\partial \eta} & 0 & \dfrac{\partial N_k}{\partial \eta} & 0 & \dfrac{\partial N_l}{\partial \eta} & 0 \end{bmatrix} \begin{bmatrix} u_i \\ v_i \\ u_j \\ v_j \\ u_k \\ v_k \\ u_l \\ v_l \end{bmatrix}.
$$

The derivatives of the shape functions, with respect to the natural co-ordinates, are available from Equations (7.9).

$$
\begin{bmatrix} \dfrac{\partial u}{\partial x} \\[2mm] \dfrac{\partial u}{\partial y} \end{bmatrix} = [J]^{-1} \begin{bmatrix} \dfrac{\partial u}{\partial \xi} \\[2mm] \dfrac{\partial u}{\partial \eta} \end{bmatrix} = [J]^{-1} \begin{bmatrix} -\dfrac{(1-\eta)}{4} & 0 & \dfrac{(1-\eta)}{4} & 0 & \dfrac{(1+\eta)}{4} & 0 & -\dfrac{(1+\eta)}{4} & 0 \\[3mm] -\dfrac{(1-\xi)}{4} & 0 & -\dfrac{(1+\xi)}{4} & 0 & \dfrac{(1+\xi)}{4} & 0 & \dfrac{(1-\xi)}{4} & 0 \end{bmatrix} \begin{bmatrix} u_i \\ v_i \\ u_j \\ v_j \\ u_k \\ v_k \\ u_l \\ v_l \end{bmatrix}.
$$

$$
\tag{7.13}
$$

The derivatives of v can be written down in a similar manner:

$$
\begin{bmatrix} \dfrac{\partial v}{\partial x} \\[2mm] \dfrac{\partial v}{\partial y} \end{bmatrix} = [J]^{-1} \begin{bmatrix} \dfrac{\partial v}{\partial \xi} \\[2mm] \dfrac{\partial v}{\partial \eta} \end{bmatrix} = [J]^{-1} \begin{bmatrix} 0 & -\dfrac{(1-\eta)}{4} & 0 & \dfrac{(1-\eta)}{4} & 0 & \dfrac{(1+\eta)}{4} & 0 & -\dfrac{(1+\eta)}{4} \\[3mm] 0 & -\dfrac{(1-\xi)}{4} & 0 & -\dfrac{(1+\xi)}{4} & 0 & \dfrac{(1+\xi)}{4} & 0 & \dfrac{(1-\xi)}{4} \end{bmatrix} \begin{bmatrix} u_i \\ v_i \\ u_j \\ v_j \\ u_k \\ v_k \\ u_l \\ v_l \end{bmatrix}.
$$

$$(7.14)$$

The strain vector can be written down from Equations (7.13) and (7.14), forming the [B] matrix as appropriate.

$$
\begin{bmatrix} \varepsilon_x \\ \varepsilon_y \\ \gamma_{xy} \end{bmatrix} = \begin{bmatrix} \dfrac{\partial u}{\partial x} \\[2mm] \dfrac{\partial v}{\partial y} \\[2mm] \dfrac{\partial u}{\partial y} + \dfrac{\partial v}{\partial x} \end{bmatrix} = [B] \begin{bmatrix} u_i \\ v_i \\ u_j \\ v_j \\ u_k \\ v_k \\ u_l \\ v_l \end{bmatrix}
$$

$$(7.14.1)$$

Note that the [B] matrix has three rows and eight columns. Each of the entries in the first row of [B] is formed by multiplying the first row of $[J]^{-1}$ by a column of the derivative matrix shown in Equation (7.13) and each of the entries in the second row of [B] is formed by multiplying the second row of $[J]^{-1}$ by a column of the derivative matrix shown in Equation (7.14). The entries in the third row of [B] have contributions from both of Equations (7.13) and (7.14)—the second row from Equation (7.13) and the first row from Equation (7.14).

The [B] matrix is, thus, dependent on the inverse of the Jacobian, which varies from point to point in the element but can be evaluated at any point, and also the derivative matrices of Equations (7.13) and (7.14) which themselves vary with position within the element.

Now the stiffness matrix can be found from Equation (4.11.1). Assuming that the element is of constant thickness (t), the integration $dVol$ can be replaced by $t\,dA$.

$$[k]_{elem} = \int_{Area} [B]^T [D][B]\, t\, dA$$

It is convenient, in this case, to integrate with respect to the natural co-ordinates rather than the Cartesian ones.

It can readily be shown that

$$dA = dxdy = \det J\, d\xi\, d\eta$$

where $\det J$ is the determinant of the Jacobian matrix.

Then

$$[k]_{elem} = \iint_{Area} [B]^T [D][B]\, t\, \det[J]\, d\xi\, d\eta \tag{7.15}$$

The stiffness matrix will, in practice, be evaluated by numerical integration of Equation (7.15).

7.2.3 A special case—the rectangular element

For the special case of a rectangular element of side length r in the x-direction and s in the y-direction, the entries in the Jacobian are constants.

$$[J] = \begin{bmatrix} \dfrac{\partial x}{\partial \xi} & \dfrac{\partial y}{\partial \xi} \\ \dfrac{\partial x}{\partial \eta} & \dfrac{\partial y}{\partial \eta} \end{bmatrix} = \frac{1}{2}\begin{bmatrix} r & 0 \\ 0 & s \end{bmatrix} \qquad : \qquad [J]^{-1} = \frac{2}{rs}\begin{bmatrix} s & 0 \\ 0 & r \end{bmatrix}$$

Evaluating [B] from Equations (7.13) and (7.14):

$$[B] = \frac{1}{2rs}\begin{bmatrix} -s(1-\eta) & 0 & s(1-\eta) & 0 & s(1+\eta) & 0 & -s(1+\eta) & 0 \\ 0 & -r(1-\xi) & 0 & -r(1+\xi) & 0 & r(1+\xi) & 0 & r(1-\xi) \\ -r(1-\xi) & -s(1-\eta) & -r(1+\xi) & s(1-\eta) & r(1+\xi) & s(1+\eta) & r(1-\xi) & -s(1+\eta) \end{bmatrix}$$

$$\tag{7.16}$$

Substituting $\xi = \dfrac{2}{r}\left(x - \dfrac{(x_i + x_k)}{2}\right)$ and $\eta = \dfrac{2}{s}\left(y - \dfrac{(y_i + y_k)}{2}\right)$ into the above equation will yield.

Equation (4.21), which was developed without recourse to the natural co-ordinates.

For this case, with the constant Jacobian, the integration of Equation (7.16) can be carried out analytically. For plane stress, the [D] matrix contains only constants but the [B] matrix contains terms that are dependent on the ξ and η co-ordinates.

The matrix [B]^T[D] has eight rows and three columns. For plane stress (using Equation (4.4a)):

$$[B]^T[D] = \frac{E}{2rs(1-v^2)}\begin{bmatrix} -s(1-\eta) & -vs(1-\eta) & -r\dfrac{(1-v)}{2}(1-\xi) \\ \cdot & \cdot & \cdot \\ \cdot & \cdot & \cdot \\ \cdot & \cdot & \cdot \\ \cdot & \cdot & \cdot \\ \cdot & \cdot & \cdot \\ \cdot & \cdot & \cdot \\ \cdot & \cdot & \cdot \end{bmatrix} \cdot$$

$$[B]^T[D][B] = \frac{E}{4r^2s^2(1-v^2)}\begin{bmatrix} s^2(1-\eta)^2 + \dfrac{(1-v)}{2}r^2(1-\xi)^2 & \cdot & \cdot & \cdot & \cdot & \cdot & \cdot & \cdot \\ \cdot & & \cdot & \cdot & \cdot & \cdot & \cdot & \cdot \\ \cdot & & \cdot & \cdot & \cdot & \cdot & \cdot & \cdot \\ \cdot & & \cdot & \cdot & \cdot & \cdot & \cdot & \cdot \\ \cdot & & \cdot & \cdot & \cdot & \cdot & \cdot & \cdot \\ \cdot & & \cdot & \cdot & \cdot & \cdot & \cdot & \cdot \\ \cdot & & \cdot & \cdot & \cdot & \cdot & \cdot & \cdot \end{bmatrix}$$

Hence the first term in the stiffness matrix for the element is:

$$k_{11} = \int\limits_{-1}^{1}\int\limits_{-1}^{1} \frac{E}{4r^2s^2\left(1-v^2\right)}\left[s^2\left(1-\eta\right)^2 + \frac{\left(1-v\right)}{2}r^2\left(1-\xi\right)^2 \right] t\ \det[J]\ d\xi\ d\eta$$

For the rectangular element det [J] is rs/4, and:

$$k_{11} = \frac{Et}{16rs\left(1-v^2\right)}\int\limits_{-1}^{1}\int\limits_{-1}^{1}\left(s^2\left(1-\eta\right)^2 + \frac{\left(1-v\right)}{2}r^2\left(1-\xi\right)^2 \right)d\xi\ d\eta$$

$$= \frac{Et}{16rs\left(1-v^2\right)}\int\limits_{-1}^{1}\left[\left(s^2\left(1-\eta\right)^2\xi - \frac{\left(1-v\right)}{6}r^2\left(1-\xi\right)^3 \right)\right]_{-1}^{1} . \ d\eta$$

$$= \frac{Et}{16rs\left(1-v^2\right)}\int\limits_{-1}^{1}\left(2s^2\left(1-\eta\right)^2 + \frac{4\left(1-v\right)}{3}r^2 \right)d\eta$$

$$= \frac{Et}{16rs\left(1-v^2\right)}\left[-\frac{2}{3}s^2\left(1-\eta\right)^3 + \frac{4\left(1-v\right)}{3}r^2\eta \right]_{-1}^{1}$$

$$= \frac{Et}{16rs\left(1-v^2\right)}\left(\frac{16}{3}s^2 + \frac{8\left(1-v\right)}{3}r^2 \right)$$

$$\therefore\quad k_{11} = \frac{Et}{6rs\left(1-v^2\right)}\left(2s^2 + \left(1-v\right)r^2 \right)$$

This expression for k_{11} is, of course, identical to that developed in Section 4.4 for the same quadrilateral. In both cases, the stiffness term has been evaluated using exact analytical integration.

The usefulness and generality of the parametric element formulation should now be apparent. The quadrilateral element developed in this section is not restricted to a rectangular shape. This means that it will be possible to pave any area with parametric quadrilaterals. The stiffness matrix of each will be evaluable based on the element formulation in parametric space and the Jacobian matrix relating its co-ordinates in parametric space to those in real space. There are, however, restrictions on the degree of distortion that will be acceptable. The variation of the Jacobian matrix over the element will be one measure of the distortion of the element.

7.3 Isoparametric Eight-Node Quadrilateral Element

The four-node quadrilateral is an essentially linear displacement element and, since strain is the first derivative of displacement, the strain field is fundamentally constant. (The actual distribution of strain and stress that can be modelled by the standard four-node quadrilateral has been discussed in Section 4.4.3.) A large number of elements will be required to approximate a rapidly-varying stress field.

The question arises as to whether a finite element can be developed to model more complex displacement and stress fields. The parametric approach is ideal for the formulation of higher-order elements. The logical first step is to introduce extra nodes on the boundaries of the quadrilateral so that the displacement along an edge can be represented by a quadratic equation. This will have the added benefit that the edge itself can now be curved, and a more accurate representation of curved boundaries will be available.

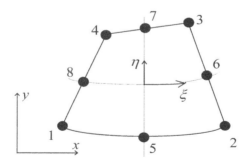

Figure 7.4 Isoparametric eight-node quadrilateral element.

137

7.3.1 *Shape functions*

As before, assume that the displacements anywhere in the element can be expressed as functions of the displacements of the nodes. The functions must satisfy the boundary conditions, $u = u_i$ and $v = v_i$ at node i. For the sake of clarity, numbers rather than letters are used for the specific node labels in this section, and the letter i is used as a general node label. In this case, functions of the form

$$u = N_1 \, u_1 + N_2 \, u_2 + N_3 \, u_3 + N_4 \, u_4 + N_5 \, u_5 + N_6 \, u_6 + N_7 \, u_7 + N_8 \, u_8 = \sum_{i=1}^{8} N_i \, u_i$$

(7.16a)

$$v = N_1 \, v_1 + N_2 \, v_2 + N_3 \, v_3 + N_4 \, v_4 + N_5 \, v_5 + N_6 \, v_6 + N_7 \, v_7 + N_8 \, v_8 = \sum_{i=1}^{8} N_i \, v_i$$

(7.16b)

are chosen, in which the shape function N_i adopts a value of one at $x = x_i$ and of zero at all other nodes. Once again, a mapping function is required, relating points in parametric space to points in Cartesian space, and for the isoparametric formulation the same shape functions are chosen.

$$x = N_1 \, x_1 + N_2 \, x_2 + N_3 \, x_3 + N_4 \, x_4 + N_5 \, x_5 + N_6 \, x_6 + N_7 \, x_7 + N_8 \, x_8 = \sum_{i=1}^{8} N_i \, x_i$$

(7.17a)

$$y = N_1 \, y_1 + N_2 \, y_2 + N_3 \, y_3 + N_4 \, y_4 + N_5 \, y_5 + N_6 \, y_6 + N_7 \, y_7 + N_8 \, y_8 = \sum_{i=1}^{8} N_i \, y_i$$

(7.17b)

The shape functions are no longer linear, but now contain quadratic terms. The shape function for a corner node is illustrated below.

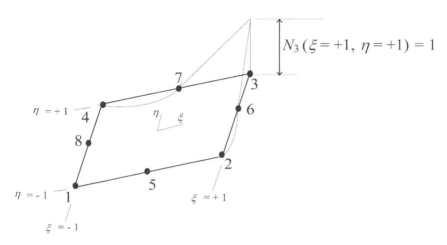

Figure 7.5 Isoparametric eight-node quadrilateral element shape function.

It can be shown that shape functions for the corner nodes are as follows.

$$N_4 = -\frac{1}{4}(1-\xi)(1+\eta)(1+\xi-\eta) \qquad N_3 = -\frac{1}{4}(1+\xi)(1+\eta)(1-\xi-\eta)$$

$$N_1 = -\frac{1}{4}(1-\xi)(1-\eta)(1+\xi+\eta) \qquad N_2 = -\frac{1}{4}(1+\xi)(1-\eta)(1-\xi+\eta)$$

$$(7.18a)$$

For the mid-side nodes:

$$N_7 = \frac{1}{2}\left((1-\xi^2)(1+\eta)\right)$$

$$N_8 = \frac{1}{2}\left((1-\xi)(1-\eta^2)\right) \qquad N_6 = \frac{1}{2}\left((1+\xi)(1-\eta^2)\right). \quad (7.18b)$$

$$N_5 = \frac{1}{2}\left((1-\xi^2)(1-\eta)\right)$$

An alternative expression for the shape functions for the corner nodes is presented below. This form illustrates how they can be built up from the shape functions of the four-node element, together with the mid-side node shape functions.

Alternative expression of corner node shape functions:

$$N_4 = \frac{1}{4}\left((1-\xi)(1+\eta)\right) - \frac{1}{2}(N_7 + N_8) \qquad N_3 = \frac{1}{4}\left((1+\xi)(1+\eta)\right) - \frac{1}{2}(N_6 + N_7)$$

$$N_1 = \frac{1}{4}\left((1-\xi)(1-\eta)\right) - \frac{1}{2}(N_8 + N_5) \qquad N_2 = \frac{1}{4}\left((1+\xi)(1-\eta)\right) - \frac{1}{2}(N_5 + N_6)$$

$$(7.18c)$$

7.3.2 The element stiffness matrix

The stiffness matrix can be derived in the same way as the one for the four-node quadrilateral. The 2×2 Jacobian matrix can be assembled by differentiation of Equations (7.17) with respect to the natural co-ordinates, substituting the shape functions from Equation (7.18). In the general case, each entry in the Jacobian is a quadratic polynomial in the natural co-ordinates.

$$j_{ij} = f(1, \xi, \eta, \xi^2, \xi\eta, \eta^2)$$

The Jacobian can be evaluated at any point, together with its inverse if it exists.

The derivatives of u and v can be calculated by substituting Equations (7.16) into Equation (7.12), and differentiating the shape functions presented in Equation (7.18). The [B] matrix, containing derivatives of the shape functions, can then be assembled as before. Note that the [B] matrix now has three rows and sixteen columns (one for each of the degrees of freedom of the elements).

The analytical evaluation of the integrals required to form the stiffness matrix is now a long and tedious exercise, if it is possible at all. The student is invited to attempt to derive an analytical expression for k_{11}, following the methodology used for the four-node quadrilateral. In practice, the time has come to investigate alternative ways to evaluate the integrals, and numerical integration is the clear answer.

7.3.3 *The general load vector: nodal equivalent loads*

Recall that the general load vector contains terms representing the influences of body forces and surface tractions.

$$\{F\} = \sum_{elems} \left(\int_{Vol} [N]^T \{W\} \, dVol + \int_{Surf} [N]^T \{p\} \, dSurf \right) + \{X\}$$

For the three-node triangle, it was shown that one half of the total load applied as a distributed pressure should be applied to each of the nodes at the ends of the side. This distribution of load is intuitively obvious. Consider now the case of a pressure load on one side of an eight-node quadrilateral element. For simplicity, it will be assumed that the quadrilateral is rectangular and has its edges parallel to the global axes.

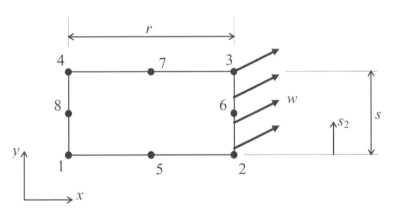

Figure 7.6 Eight-node quadrilateral element with pressure load.

Assume that the element has a line load per unit length with components w_x and w_y applied to the right hand side (side 2-6-3) of the element and that the length of the side is s. The pressure on the surface has components p_x and p_y, calculated by dividing the line load by the element thickness.

$$\{X\}_{surf} = \int_{Surf} [N]^T \{p\} \, dSurf = \int_{2\text{-}6\text{-}3} \begin{bmatrix} N_1 & 0 \\ 0 & N_1 \\ \vdots & \vdots \\ \vdots & \vdots \\ N_8 & 0 \\ 0 & N_8 \end{bmatrix} t \, ds_2 \begin{bmatrix} p_x \\ p_y \end{bmatrix}$$

Now the integral over the side 2-6-3 of each of the shape functions is required.

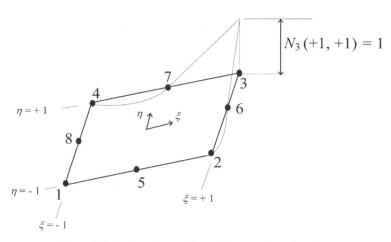

$N_3\,(+1, +1) = 1$

Figure 7.7 Eight-node quadrilateral element shape function.

The integral of the shape function associated with the corner node 3 over the side 2-6-3 is:

$$\int_{2\text{-}6\text{-}3} N_3 \, .dSurf \;=\; \int_0^s -\frac{1}{4}(1+\xi)(1+\eta)(1-\xi-\eta).t.ds_2$$

$$=\; \int_{-1}^1 -\frac{1}{4}(1+1)(1+\eta)(1-1-\eta).t.\frac{s}{2}.d\eta$$

$$= \frac{st}{4}\int_{-1}^{1}\left(\eta+\eta^{2}\right).d\eta = \frac{st}{4}\left[\frac{\eta^{2}}{2}+\frac{\eta^{3}}{3}\right]_{-1}^{+1}$$

$$= \frac{st}{6}$$

Since the shape function associated with node 2 is of the same shape it must have the same integral for this undistorted element.

Thus

$$\int_{2-6-3} N_{2}\ dSurf = \frac{st}{6}$$

The integral of the shape function associated with the midside node 6 over the side 2-6-3 is:

$$\int_{2-6-3} N_{6}\ dSurf = \int_{0}^{s}\frac{1}{2}(1+\xi)\left(1-\eta^{2}\right).t.ds_{2}$$

$$= \int_{-1}^{1}\frac{1}{2}(1+1)\left(1-\eta^{2}\right).t.\frac{s}{2}.d\eta$$

$$= \frac{st}{2}\int_{-1}^{1}\left(1-\eta^{2}\right).d\eta = \frac{st}{4}\left[\eta-\frac{\eta^{3}}{3}\right]_{-1}^{+1}$$

$$= \frac{4st}{6}$$

The shape function N_{1} is zero at all points along the line 2-6-3 so:

$$\int_{2-6-3} N_{1}\ dSurf = 0$$

The same is true of the shape functions associated with the remaining nodes (4, 5, 7 and 8), i.e.:

$$\int_{2-6-3} N_4 \; dSurf = \int_{2-6-3} N_5 \; dSurf = \int_{2-6-3} N_7 \; dSurf = \int_{2-6-3} N_8 \; dSurf = 0$$

Thus the element force vector consistent with the pressure on the side 2-6-3 is:

$$\{X\}_{surf} = \int_{2-6-3} \begin{bmatrix} N_1 & 0 \\ 0 & N_1 \\ N_2 & 0 \\ 0 & N_2 \\ N_3 & 0 \\ 0 & N_3 \\ N_4 & 0 \\ 0 & N_4 \\ N_5 & 0 \\ 0 & N_5 \\ N_6 & 0 \\ 0 & N_6 \\ N_7 & 0 \\ 0 & N_7 \\ N_8 & 0 \\ 0 & N_8 \end{bmatrix} .t.dSurf \begin{bmatrix} p_x \\ p_y \end{bmatrix} = \frac{ts}{6} \begin{bmatrix} 0 \\ 0 \\ p_x \\ p_y \\ p_x \\ p_y \\ 0 \\ 0 \\ 0 \\ 0 \\ 4p_x \\ 4p_y \\ 0 \\ 0 \\ 0 \\ 0 \end{bmatrix}$$

Hence, the appropriate treatment of a uniform pressure force on the side of the element is to distribute one-sixth to each of the corner nodes and four-sixths to the midside node, i.e., a 1:4:1 distribution of the total load.

This distribution of load is not obvious, and, furthermore, it will change if the element is distorted. In the general case, the Jacobian matrix is used to express the integral over the edge of the element in parametric co-ordinates, and the required distribution of load is dependent on the variation of the Jacobian along the edge.

It will clearly be better for the calculation of equivalent loads to be done by the analysis program and not by the analyst in the general case.

Suggested Reading

Adams, V. and Askenazi, A. 1999. Building Better Products with Finite Element Analysis, OnWord Press.

Akin, J. E. 1994. Finite Elements for Analysis and Design, Academic Press.

Akin, J. E. 2005. Finite Element Analysis with Error Estimators: An Introduction to the FEM and Adaptive Error Analysis for Engineering Students, Elsevier Science.

Akin, J. E. 2010. Finite Element Analysis Concepts: Via SolidWorks, World Scientific.

Alvala, C. R. 2008. Finite Element Methods: Basic Concepts and Applications, PHI Learning.

Baaser, H. 2010. Development and Application of the Finite Element Method based on MATLAB®, Springer Berlin Heidelberg.

Babuska, I., Whiteman, J. and Strouboulis, T. 2010. Finite Elements: An Introduction to the Method and Error Estimation, OUP Oxford.

Barbero, E. J. 2013. Finite Element Analysis of Composite Materials Using ANSYS®, Second Edition, Taylor & Francis.

Bathe, K. J. 2006. Finite Element Procedures, Prentice Hall.

Belytschko, T., Liu, W. K., Moran, B. and Elkhodary, K. 2014. Nonlinear Finite Elements for Continua and Structures, Wiley.

Bhatti, M. A. 2005. Fundamental Finite Element Analysis and Applications: with Mathematica and MATLAB® Computations, Wiley.

Bi, Z. 2017. Finite Element Analysis Applications: A Systematic and Practical Approach, Elsevier Science.

Buchanan, G. R. 1995. Schaum's Outline of Finite Element Analysis, McGraw-Hill Education.

Champion, E. R. 1988. Finite Element Analysis with Personal Computers, Taylor & Francis.

Chandrupatla, T. R. and Belegundu, A. D. 2014. Introduction to Finite Elements in Engineering: International Edition, Pearson Education Limited.

Chen, X. and Liu, Y. 2014. Finite Element Modeling and Simulation with ANSYS Workbench, Taylor & Francis.

Cook, R. D. 2001. Concepts and Applications of Finite Element Analysis, Wiley.

De Borst, R. Ã., Crisfield, M. A., Remmers, J. J. C. and Verhoosel, C. V. 2012. Nonlinear Finite Element Analysis of Solids and Structures, Wiley.

Desai, C. S. and Kundu, T. 2001. Introductory Finite Element Method, Taylor & Francis.

Desai, Y. M. 2011. Finite Element Method with Applications in Engineering, Dorling Kindersley.

Dill, E. H. 2011. The Finite Element Method for Mechanics of Solids with ANSYS Applications, CRC Press.

Donald, B. J. M. 2011. Practical Stress Analysis with Finite Elements, Glasnevin Publishing.

Doyle, J. F. 2014. Nonlinear Structural Dynamics using FE Methods, Cambridge University Press.

Ellobody, E., Feng, R. and Young, B. 2013. Finite Element Analysis and Design of Metal Structures, Elsevier Science.

Ferreira, A. J. M. 2008. MATLAB® Codes for Finite Element Analysis: Solids and Structures, Springer Netherlands.

Fish, J. and Belytschko, T. 2007. A First Course in Finite Elements, Wiley.

Gokhale, N. S. 2008. Practical Finite Element Analysis, Finite to Infinite.

Gosz, M. R. 2017. Finite Element Method: Applications in Solids, Structures, and Heat Transfer, CRC Press.

Huebner, K. H. 2001. The Finite Element Method for Engineers, Wiley.

Hughes, T. J. R. 2000. The Finite Element Method: Linear Static and Dynamic Finite Element Analysis, Dover Publications.

Hughes, T. J. R. 2012. The Finite Element Method: Linear Static and Dynamic Finite Element Analysis, Dover Publications.

Hutton, D. V. 2003. Fundamentals of Finite Element Analysis, McGraw-Hill.

Jin, J. M. and Riley, D. J. 2009. Finite Element Analysis of Antennas and Arrays, Wiley.

Kattan, P. I. 2014. MATLAB® Guide to Finite Elements: An Interactive Approach, Springer Berlin Heidelberg.

Khennane, A. 2013. Introduction to Finite Element Analysis using MATLAB® and Abaqus, CRC Press.

Khoei, A. R. 2015. Extended Finite Element Method: Theory and Applications, Wiley.

Kim, M. 2013. Finite Element Methods with Programming and Ansys, LULU Press.

Kim, N. H. and Sankar, B. V. 2008. Introduction to Finite Element Analysis and Design, Wiley.

King, R. H. 2018. Finite Element Analysis with SOLIDWORKS Simulation, Cengage Learning.

Kobayashi, S., Oh, S. I. and Altan, T. 1989. Metal Forming and the Finite-Element Method, Oxford University Press.

Koutromanos, I. 2018. Fundamentals of Finite Element Analysis: Linear Finite Element Analysis, Wiley.

Kwon, Y. W. and Bang, H. 2000. The Finite Element Method using MATLAB®, Second Edition, CRC Press.

Larson, M. G. and Bengzon, F. 2013. The Finite Element Method: Theory, Implementation, and Applications, Springer Berlin Heidelberg.

Liu, G. R. and Quek, S. S. 2013. The Finite Element Method: A Practical Course, Elsevier Science.

Logan, D. L. 2011. A First Course in the Finite Element Method, Cengage Learning.

Madenci, E. and Guven, I. 2015. The Finite Element Method and Applications in Engineering Using ANSYS®, Springer US.

Moaveni, S. 2015. Finite Element Analysis: Theory and Application with ANSYS, Global Edition, Pearson Education Limited.

Mohammadi, S. 2008. Extended Finite Element Method: For Fracture Analysis of Structures, Wiley.

Musa, S. M., Kulkarni, A. V. and Havanur, V. K. 2013. Finite Element Analysis: A Primer, Mercury Learning & Information.

Pavlou, D. G. 2015. Essentials of the Finite Element Method: For Mechanical and Structural Engineers, Elsevier Science.

Petyt, M. 2010. Introduction to Finite Element Vibration Analysis, Cambridge University Press.

Pidaparti, R. M. 2017. Engineering Finite Element Analysis, Morgan & Claypool Publishers.

Ragab, S. and Fayed, H. E. 2017. Introduction to Finite Element Analysis for Engineers, Taylor & Francis, CRC Press.

Rao, S. S. 2017. The Finite Element Method in Engineering, Elsevier Science.

Raoufi, C. 2017. Applied Finite Element Analysis with SolidWorks Simulation 2017.

Reddy, J. N. 1993. An Introduction to the Finite Element Method, McGraw-Hill.

Seshu, P. 2003. Textbook of Finite Element Analysis, PHI Learning.

Smith, I. M., Griffiths, D. V. and Margetts, L. 2013. Programming the Finite Element Method, Wiley.

Thompson, M. K. and Thompson, J. M. 2017. ANSYS Mechanical APDL for Finite Element Analysis, Elsevier Science.

Wriggers, P. 2010. Nonlinear Finite Element Methods, Springer Berlin Heidelberg.

Zienkiewicz, O. C. and Morgan, K. 2006. Finite Elements and Approximation, Dover Publications.

Zienkiewicz, O. C. and Taylor, R. L. 2013. The Finite Element Method for Solid and Structural Mechanics, Elsevier Science.

Index